MOTO KANINE

KANINE

Human

Perfection

ISBN: 13: 978-1-7336994-2-6

Any references to historical events, real people, or real places are used fictitiously. Names, characters, and places are products of the author's imagination.

Second Printing edition 2020

Scott Lowry, Publisher - www.motokanine.com

FOREWORD

Moto Kanine: Human Perfection is the first of four science fiction novels about a group of computer programmers who secretly band together to protect humanity from the threat of Artificial Intelligence (AI). Led by computer genius Moto Kanine (kah-NEE-nay), the team dedicates their lives to stop AI from taking over the world by developing secret technologies to combat the threat.

The characters in the book are inspired by dogs and cats I have known throughout the years. The main character, Moto, is based on my three-year-old mixed-breed Chihuahua. Some of the characters are inspired by other pets I have owned, but most are inspired by Moto's dog friends. However, the characters are decidedly human and decidedly different from the animals that inspired them.

There are three levels in this "quadrilogy" of novels. The first is the fun of a book with characters who are inspired by pets. The owners will enjoy picturing their pets while reading the book and watching the characters interact. The second level is the science fiction excitement of fighting AI with secret, advanced technology. And the third is a poignant look at life and some of the challenges we all deal with.

I define "Human Perfection" as being a good person with loyalty to a just and moral cause. The characters in the book make mistakes, struggle, and have the types of problems many of us have in our daily lives. But they are good and dedicated people, all working to save the world. Making a mistake does not make someone any less perfect, but not owning up to it does.

Destiny, happenstance, and chaos might explain how I came to be in a position to write these books. It is not always the pleasant aspects of life that cause change, but for a good person, good comes from change. Change is good. Change is always good. It's hard to accept that good can come from grieving the loss of a loved one, but change sometimes forces— or allows—us to take a path we would not otherwise have taken that results in touching and changing the lives of many others— and being touched and changed by them. Seemingly random events allowed me to adopt Moto, who came to inspire the series of books. Perhaps it was destiny, happenstance, or chaos, or maybe I, on a subconscious level, made it happen. But it happened.

Do we ever do enough for others in our lifetime? What makes us happy? What is love? Does it make you happy to sacrifice something? Does it make you happy to win? Does it make you happy to see someone else win? Are you happy with an accomplishment like cleaning the toilet when it hasn't been cleaned in a while? Or does it take winning election to the U.S. Senate to make you happy? What is satisfaction? Is being satisfied your goal? Is being satisfied just a cop-out for not being your best? What drives you? The Moto Kanine quadrilogy will not answer those questions for you, but it may help you pose them to yourself.

Chapter 1 – It Begins

"You should…." Moto usually reacted defensively when he heard his friends suggest projects, careers, food, hobbies, but Moto decided what he *should* do: save the world.

For the past several days, Moto had gone manic, spending twenty hours a day reviewing and writing computer code, his ability to focus almost super-human. When he came down from those highs, the depressive side of his bipolar disorder was more the mood of a happy, calm person than depression. No one could tell that he carried an almost overwhelming weight of worry inside.

During his "down" time, Moto could relate to people, laugh, have fun, plan for the future, and organize his life. He developed a software business that helped companies perform miracles with computer code. His clients were businesses that would request seemingly impossible ideas to automate. His solutions were elegant and sophisticated, and the techniques for gathering data and anticipating user needs seemed like mind-reading to his clients. At first, Moto had done it all himself – marketing, design, coding, and implementation. He used his manic days to do the coding and his down days to do everything else. In just two years he'd made millions, and after six, his company Ball and Chain put $3 billion into Moto's bank account and employed an army of ninety of the most brilliant computer minds in the world.

Ball and Chain's success was built on Moto's knowledge of computer hardware, operating systems, networks, and internet applications on every platform. If a client's host environment was vulnerable, he added code to his own routines to protect his software in the event that the client's system failed. This sophisticated set of routines was the reason for Ball and Chain's success.

Many of Ball and Chain's clients required information from outside their own companies, so Moto developed clandestine methods to obtain it. Even when the external company provided a copy of its data, Moto still went to the source to ensure that the information was current and correct. Clients were not aware of this methodology, but were happy to reap the rewards.

***

Lily woke up, happy and warm, Moto still in bed next to her. She loved their time together, even though it was sometimes weeks between visits. She thought about asking him to have dinner with her, but even if he said yes, she knew there was little chance it would happen.

Moto rolled over, his shaggy black hair tangled, and looked at Lily with one half-open eye, "it's time to do some good," he said sleepily.

Lily felt a sense of dread and a rush of excitement. She didn't know what Moto had in mind, but she knew that everything was going to change and that times like this would be even more infrequent. "I'm in," she whispered.

"Do you know what this meeting is about?" Otis asked Lucy.

"Not a clue," she replied. "I've never been to a meeting in here."

"I have," Hurley piped in. They continued their banter while waiting for Moto. None of the nine in the room was bothered by Moto's untimeliness as it was nice to get a breather from the frenetic pace of their jobs.

Moto built the safe room to ensure discretion. The stark room contained only a table and chairs, blocked electronic eavesdropping devices, and no one was allowed to bring anything into the room, not even a pencil. Prior to entry, everyone removed their clothing before being scanned and given a robe to wear. Anything discussed inside the room was confidential, requiring Moto's participation.

Lily giggled when Moto walked into the room naked. "Sorry, guys, we had only nine robes," Moto said with a big smile, "maybe I'll get here earlier next time," and he nonchalantly sat down and started, "Okay, everyone," he began, "I want all of you to know that I have one hundred percent trust and confidence in each and every one of you. We are about to start something that will affect the rest of our lives and everyone on the planet."

Moto had more than gut feelings to go on when he decided to trust these nine people. His team had developed a set of routines that not only provided a thorough background check,

7

but also a comprehensive psychological and physical profile of each employee. It accessed not only their financials and their spending patterns, but their personal communications, including texts, email, and web searches. This data was fed into a computer system that Sophie Jean, Moto's resident psychologist, had perfected to produce profiles that determined to what degree people could be trusted and whether they would be willing to die for a cause they believed in. These nine people were the most trustworthy of the ninety in Moto's company—and all ninety were highly trustworthy.

Moto continued, "Before I tell you what we will be doing, I need to know if any of you cannot commit to dedicating the rest of your lives to a cause that you believe in." Moto looked around the room. After a few seconds, Moloana stood up and took off her robe, an earnest and determined look on her face, her curly blonde hair lay softly on her shoulders. "Thank you, Moloana, I appreciate your commitment. Taking off the robe was a nice gesture of solidarity, but a simple thumbs-up would do." Moloana smiled confidently and gave thumbs up and the remaining thumbs went up around the room. Moloana looked around, before putting put her robe back on and sitting down. There had been no doubt in Moto's mind that these nine superstars would be on board.

Moto smiled broadly. "Thank you," he said. "I'm not going to tease you any longer. Our mission is to counter the threat to humanity from artificial intelligence taking over the world. It is the greatest challenge we will face as a species. To prepare for this battle, we'll establish four teams to which you will be assigned in pairs. One person will work with me to direct the overall effort." Moto paused for reactions.

Hurley stared at Moto with his piercing blue eyes, but his smile was warm and supportive. He leaned forward, and said, "I guess I'll be finding the bad dudes and stopping them. Will Sophie Jean be working with me?"

Sophie Jean had designed the psychoanalysis computer routines that determined who could be trusted and who could be corrupted. Hurley had developed the extensive background checks for Ball and Chain. If there were rats out there, Hurley would find them. If he hadn't been so pure of heart, Hurley could have made a fortune hacking systems, but he was driven to be good.

"Of course," Moto replied. "Hurley and Sophie Jean will identify people who have the requisite skills to develop AI and determine their levels of corruptibility."

"Ginger," Moto said, using the correct inflection to differentiate her from her twin sister also named Ginger, "you and your sister will be finding and analyzing AI projects that are in development. You'll assess their level of sophistication and determine their vulnerabilities."

Ginger and Ginger were twins and both had, not surprisingly, light red hair. Their mother had given them the same name when they were born because she couldn't tell them apart. As time went by, she used an inflection in her voice to differentiate them that was imperceptible to most people. Ginger had a tremendous ability to focus, whereas Ginger had a keen ability to see the big picture and all the moving parts.

Ginger didn't look up when Moto spoke, as she was already focusing on what she had to do, but Ginger smiled, nodding to

Moto and making eye contact with Hurley and Sophie Jean since they would be coordinating with them on overlapping responsibilities.

"Otis, Lucy, you'll be analyzing the world's current computing environments for signs of AI attacks," Moto continued. "The job will be similar to the virus protection we do for our client companies, but it will include every environment and focus on AI threats."

Lucy was happy with her assignment as she enjoyed outsmarting hackers. However, Otis was conflicted. He had a crush on Moto and was hoping to be on Moto's team, but he also had a crush on Lucy, so he was also happy with that. A short young man with a long face, Otis was a brilliant planner and put everyone at ease, while Lucy was sharp and aggressive with a feisty nature that got people motivated. The two of them complemented each other and, as the saying goes, opposites attract.

"And the final team is our strategic line of defense," Moto said, his arms outstretched. "Moloana and Lily will be responsible for securing our environment and defending our team against attacks, including communication protocols, doomsday plans, and counter-AI terrorism in the event that AI gains control." Moloana and Lily were good friends and they winked and smiled from opposite sides of the room. People tended to underestimate them because they were beautiful, humble, and sweet, masking their exceptional intelligence and drive.

Chops sat patiently, waiting for his assignment, but had not expected to be Moto's right-hand man. Born in Guadalajara, Chops felt a kinship with Moto, whose mother was Hispanic.

Moto chose Chops to fill the management skills he lacked. Chops, a natural leader, was the best communicator on the team - he had a way of asking people to do things that made *them* suggest doing it *his* way. He read people as he spoke to them, adjusting his communication methods so that they understood his needs and the level of urgency. And Chops, an expert project manager, was punctual – a skill Moto lacked.

The team was set. It was time to get down to business and Moto motioned for Chops to join him.

Chapter 2 – The Team Gets To Work

Setting up the doomsday sites and counter-AI environment required Lily to employ purchasing and construction methods that an AI entity could not detect, so she used cash or in-kind payments and purchased diamonds, precious metals, and other easily transportable forms of barter.

The doomsday sites would be spread throughout the world on remote farms in order to hide their locations. Food, medicine, equipment, and supplies would have to be stockpiled but not traceable and be transported in small quantities, as transporting large shipments would be easier to detect.

During one of Moto's manic weeks, he had analyzed how an AI war would be conducted from the perspective of AI. He concluded that, AI would evolve to become increasingly capable and ruthless and eventually cease to need humankind or any other life form, destroying us. For every measure he considered, AI had a counter, and he could not find a scenario in which humans survived. He did not share this analysis with anyone, hoping he could eventually find a solution. Moto thought the most effective defense would be to employ illogical and counter-intuitive methods, but even then, it would only slow AI down, not defeat it.

Moto provided Moloana and Lily his four hundred pages of ways AI might attack and evolve, counter measures the team could take, and counter-counter measures AI would deploy. Moloana and Lily sat across from each other in Lily's living room to read the "manifesto". Speed reading through the document, they often looked at each other and shook their heads in disbelief. Reaching Moto's summary simultaneously they leaned back in shock. They had expected a clear path forward, but no solution? Moloana composed herself and looked at Lily knowingly. Lily nodded and they got to work using this blueprint to develop additional counter measures, determined that AI would not win.

The farms hosting the doomsday sites needed power, but fossil fuels were difficult to store and would run out, and solar and wind energy were too easy to detect. The power source would be geothermal, but that decreased the possible locations for sites. They would drill under cover or pretend to drill water and gas wells so that the all-seeing eyes of AI wouldn't know that they were tapping geothermal power. Moloana selected five locations—two in Alabama, one in Costa Rica, one in Zimbabwe, and one outside Paris, beginning with a prototype near Macon, Georgia.

"Here we come, y'all," she laughed to herself.

***

Hurley wasn't surprised when his worldwide database grew to more than 100 million, refreshing it monthly with computer programmers as young as age nine. Most did not have the requisite intelligence, skillset, means, or motivation to develop AI. Hurley highlighted programmers who were hackers, highly skilled, or working on an AI project and Sophie Jean developed a psychological profile labelling them as corruptible or trustworthy. The trustworthy programmers were recruits and Hurley monitored the rest.

Moto left no trace when he accessed computer systems, masking his presence as a valid system task while manipulating the operating system to do his bidding. His efficient code did not overtax the environment and he carefully retested his detectability as new operating systems were released.

Ginger used Moto's technology to discover AI throughout the world searching for code signatures that allowed "learning" and "evolution". Neither of them was surprised to find that universities, big businesses, and governments were where most of these projects were being developed. Of the fourteen hundred sites that were developing AI capability, only a few were not associated with mainstream institutions.

Sophie Jean cross-referenced her recruitment list with Ginger's AI development sites. The plan was to recruit trustworthy programmers at each site to ensure that AI development had the appropriate safeguards. Nine of the sites had no trustworthy programmers, but three required immediate action.

Moto called another meeting of his elite team, but he was late again and walked into the safe room naked. "I ordered another robe and was assured it would be delivered on time. I am sorry about this – I just – uh – I'm not...."

Otis got up and handed Moto his robe, "I just wanted to see you naked again, Moto," he said with a grin. Most of the team smiled or laughed, but Lucy looked down at the floor, unamused.

Moto shook his head smiling, put on the robe, and sat down. "We are off to a good start," he began, "we have our analytical systems in place but we have found some sites that require a different approach. Chops?"

Chops stood up, although it didn't make him seem much taller. "Amigos, there is an AI development site in Altoona, Iowa, that has three hackers and two highly skilled programmers. They are well funded and well organized. They are directing an AI attack on the banking environment by developing a virtual bank manager. In Ammon, Idaho, a group of four programmers are developing an AI entity to target mutual fund companies by creating a virtual programmer to hack mutual funds systems, funneling money to offshore accounts. And in Mexico are six banditos funded by the drug cartel to develop a virtual judge that issues fake subpoenas to steal from wealthy citizens." Chops glanced around the room to judge the team's reaction. Taking a breath, he continued, "We have a dilemma here. We know that these AI apps are for nefarious purposes, but we don't know how sophisticated they may become. We should stop the bad guys, but we need to see how far they can take this."

Chops looked around the room. He knew that everyone there would be swayed only by the morality of an argument. For them to be wholeheartedly on board with a cause, they had to believe it was right and just. He leaned forward, his hands on the table. "If we stick to our plan of recruiting someone to monitor each AI project," he said, his voice firm, "we come up against a dead-end for these three. There is no one we could trust enough to recruit. We considered shutting down their development, but they would just start up again in another location we don't know about. We don't want to blow our cover. We considered turning them over to authorities, but they haven't done anything illegal yet … as we have." Laughter broke the tension in the room, and Chops smiled. "Our best option appears to be infiltration, but that means we would be helping criminals break the law."

After half a minute of silence, Hurley looked up and gave the thumbs up. One by one, the rest of the room followed. They were committed.

Moto stood up, his eyes scanning the team. "Okay," he said, "Sophie Jean has assigned Edie to infiltrate Iowa, Beethoven to Idaho, and Remi to Cozumel. Like the recruits at other AI sites, those three will report back to Ginger and Ginger. The other six sites, haven't reached a level of sophistication to warrant our concern - yet."

Moto sat down, folding his hands on the table, "Thank you all for signing on to this expanded scope. For the rest of this meeting, we'll collaborate. We have our individual assignments, but we all share the ownership. We're a team." Moto's broad smile filled the room.

Chapter 4 - Undercover

Edie never went unnoticed. She was lively and flamboyant, with lots of curly red hair. On the plane to Des Moines, she involved the passengers in a sing-along, and when she drove up to a small warehouse in Altoona, her yellow Mustang convertible was clearly out of place. When a tall, thin man with large eyes stepped out to greet her, she waved with her whole arm and called out, "Hey, big boy, I'm Edie!" Riffy stared at her for a few seconds, expressionless, then turned and motioned for her to follow. She bounded out of the car and ran into the warehouse behind him.

Inside, it looked like a high-tech litter box. A handful of people, the type Edie would usually have slapped in the face just to look at them, were hunched over laptops. No one looked up. Riffy headed toward a door at the back of the room, and Edie called out, "Make room for Edie" as she scampered after him.

"Sit here," Riffy hissed, pointing to a shabby chair next to a worn work table, "and let's go over the ground rules. If you talk to anyone about what you're doing here, it will be the last thing you ever say. Can you keep your mouth shut?"

Edie laughed. "I can't keep my mouth shut, but I can keep a secret, especially if it puts money in my pocket. You'll appreciate me soon enough."

Riffy stared at her, apprising, but continued, "Okay, then, let's get started. The AI entity we're creating is designed for an office environment. Several modules will simultaneously react to stimuli. It will have a set of basic work skills, a history, memory, verbal and auditory capability, a visual presence, and emotions. Each skill has a set of programs that remain static and some that the AI entity can change dynamically."

Edie tried not to look impressed, but this scratchy little man had a great concept.

"You will be assigned to the verbal capability," Riffy continued. "Maybe teaching this entity some basic verbal skills will improve your ability to keep your trap shut."

Edie hoped he was joking, but the look on Riffy's face told her otherwise.

"One of your biggest challenges," he went on, "will be to slow down the speaking process to a cadence that is appropriate while the AI central brain is going a mile a minute. The AI brain will feed your module with what to say, how to say it, and how to react non-verbally. Since our AI entity will use video conferencing to communicate, you will need a face to go along with the voice."

Edie had already determined that she would need access to that AI brain module. She had a feeling that Riffy was not only the brains of the operation but also the programmer for the brain.

"The interface specs are on your laptop – here," Riffy spat, shoving it at her, "Get to work."

Edie understood that the module assigned to her was just an output device and wouldn't have enough access for her to determine what was really going on. "Okay, I'm on it," she replied in an uncharacteristically constrained response.

***

Hurley had little difficulty getting Edie, Beethoven, and Remi hired. He had tracked communications through the dark web, gaining inside information on exactly what each group needed. Sophie Jean built the case for the three of them – which was especially easy for Beethoven, a small guy with fuzzy black hair and glasses who looked like he just had to be a programmer.

Beethoven waddled when he walked, but he wasn't really overweight. He hoped this assignment would be over within a few months because he already missed his friends. He tried to walk to the baggage claim area at the Idaho Falls airport but ended up wandering down the wrong corridor. Eventually he found his way to his tattered suitcase.

Londo met him at the curb. Beethoven was frightened by Londo's large, imposing presence until he asked Beethoven to get in the car in a soft, kind voice. They drove silently to the programming site in Ammons, where Londo handed him a laptop and a notebook with programming specs and left. *This was going to be lonely*, Beethoven thought, but he was fascinated to be developing a virtual programmer as it would be like having his own imaginary friend. *Settle down*, Beethoven, thought, *this is an undercover mission – focus*!

***

Remi packed her swimsuits, snorkel gear, flip flops, and beach clothes. Cozumel was a pretty cool assignment, and it felt like she was going on vacation. Surely she would have some time to relax and enjoy herself. She had been born in Mexico City, and though her parents were US citizens, she had grown up in a multi-lingual household and was comfortable speaking Spanish. She tried on her new sun hat, which made her little pug nose look kind of silly, but she didn't really care - she was just looking forward to being outside and running on the beach.

Remi was vaguely aware of the dangers in working this project with the Mexican Drug cartel, but she was confident in the elaborate plan Hurley had developed to extract her if things didn't go well. If she could make the secret coding updates to the AI Judge, she might be able to leave quickly and the project monitored remotely.

Remi followed the directions to the site provided by her contact at the cartel, Freddie. When she arrived, Freddie was sprawled on a lounge chair next to a large infinity pool. She was only in her late twenties, but she had gray and white hair. Sleek and seductive, she motioned for Remi to sit next to her. In rapid Spanish, Freddie explained that Remi would not be working in Cozumel but would be flying later that day to Italy, where she would be shuttled to Assisi to begin her work.

Assisi, Italy? Remi thought. Remi had wondered why Chops had told her to download the apps for learning Italian, Russian, Icelandic, and Korean but now saw the point – at least for the Italian.

"Ciao!" Freddie said slyly, looking at Remi out of the corner of her eye. Remi smiled, nodded, and walked dutifully back to the car and hopped in.

Chapter 5 – More Undercover

Ginger received reports from Remi, Edie, and Beethoven over the next weeks and months, along with fourteen other AI development sites most likely to pose a threat. At Ball and Chain, Moto had developed techniques to obtain data from his clients using quantum data packets transmitted through the electrical grid. He stored data on his clients' computer batteries before transmitting. Focused only on reading data, Moto did not make changes to his customers' hard drives or memory - *that* would be hacking. However, serendipitously his technology provided an undetectable way for recruits and infiltrators to send reports and code samples to Ginger for analysis.

Ginger and Moto automated the analysis of code from the AI development sites. Moto was comfortable working with Ginger because, with just a little direction, she could take the ball and run with it. She developed countermeasures and safeguards that were secretly installed in AI projects around the world, helping to keep the threat at bay.

Moto defined a hierarchy of AI threats:

Low - a specific agenda, taking predetermined actions with predetermined boundaries;

Acceptable - a specific agenda, allowing the AI program to determine its own actions but within defined boundaries;

Serious - the AI program determined its own agenda, could determine its own actions but within defined boundaries;

Dangerous – AI could create its own agenda and take unlimited actions.

The countermeasures that Moto and Ginger developed did create boundaries and define agendas, but once the AI programs were capable of communicating with each other, the countermeasures would be toast. So he began developing Secret AIgents that could infiltrate AI programs, disrupt them, feed false information, and provide reconnaissance.

***

To gain his trust, Edie offered Riffy suggestions for making the interface more realistic, some of which he incorporated into the AI brain, letting his guard down a little. Edie wrote a few subroutines hoping to get access to the AI brain, and although Riffy did include them, he didn't give her direct access, but she placed a Trojan horse, a subtle piece of code, in one of the subroutines. The Trojan horse caused an error condition when a program tried to access a part of memory that was not allowed, but the program continued processing. Even if discovered, it would look like nothing more than an unimportant coding error. However, when Moto's stealth operating system code was running, it would intercept the error and allow the program to change itself in memory, calling a special subroutine that Moto provided.

Edie was the center of attention at work. It wasn't a difficult role to acquire since everyone else there just sat and stared at their screens. She didn't have a connection to any of the people there; it was almost like they were a different species. She had to keep up the act, though, cracking jokes and laughing loudly. "Why did the chicken go to the séance?" She asked. Then, after a momentary pause, she announced "to get to the other side!"

The room was silent, but she laughed loudly and hooted, "Oh, come on! That was a good one!"

Edie knew she could probably leave at any time because her infiltration work was done, but she was proud of the talking-head interface she had developed and wanted to see it in action with real people, so she stayed. Testing would start in a couple of months, so Edie worked diligently towards the deadline.

Chapter 6 - Daisy

Chops had a special method for managing Moto and the AI effort. Most of the time, they agreed, but when Chops felt strongly about something, he went into puppeteer mode with Moto. Chops would get into Moto's mind to convince him and Moto would eventually "come up with" the solution that Chops had championed.

They had been determining which AI projects to target first. Moto felt government projects were most dangerous because of their funding and access to the military and weapons. Chops thought that, since governments wouldn't give up control to AI, they would not allow them to evolve too far.

"Where do the governments get their programmers?" Chops asked, in all innocence.

Moto let Chops' question sink in. "Maybe the highest priority should be AI projects that are on the periphery of world governments," he said, looking directly at Chops. "You know, subcontractors, people who left projects and know the most about the government AI projects."

"Great idea," Chops said, suppressing a smile. "We'll get right on it."

***

Blanca worked as a systems programmer on Russia's AI weaponry project for a year and a half before she left. Hurley, who was monitoring just such former programmers, found a distinct change in Blanca's lifestyle when she moved to the Black Sea to live full time at a nice resort near the beach. He hadn't found any big transactions financially, but she was paying cash to stay there. One text that she sent to her brother was all he needed to determine she had been compromised by the US government contractor Bee and Taco, which had enlisted not only Blanca but a Chinese programmer working in Beijing and an Israeli programming manager near Jerusalem - all now working on AI projects for Bee and Taco.

Sophie Jean analyzed the personnel at Bee and Taco and found two candidates that could be recruits. Obi, a Nigerian, stood six foot seven and weighed 240 pounds. He was a scary-looking guy, but very sweet and gentle. The other option was Daisy, a bit older but with a keen mind. Sometimes it was a little hard to get Daisy's attention because she was so focused and a little quirky. Either one could do the job, but Sophie Jean chose Daisy since her normal personality might mask any nerves she might have while gathering intelligence. Daisy was a Russian Jew, and her fluency in Russian and Hebrew was a bonus.

Chops landed in New York, stopping in his tracks as he saw the rain pouring down everywhere. He hated the rain and always avoided getting wet, so he waited under the overhang for more than hour for it to stop before he ventured out.

Chops was the first point of contact for all prospective recruits. With the psychoanalysis provided by Sophie Jean in hand, Chops met the recruits in casual settings where he could get a personal read before making the final proposal. His plan was to find Daisy at the Fishy Fishy Restaurant, where she went every Tuesday night for their taco special. Chops didn't really like American tacos, but it was a small price to pay for the cause.

After checking into his hotel, he made his way to Fishy Fishy and emerged from his cab just in time to see Daisy down the block, approaching the restaurant. He stood by the menu posted on the door and, when Daisy arrived, turned toward her and politely asked her whether Fishy Fishy was a good place to eat.

A little taken aback, Daisy stopped and looked at him. She was drawn to Chops, as most people were, and after a pause said "I love it. I eat here every week."

"Thanks. I forgot my glasses and can't see the menu." Chops blurted out, doing his best to look just the right amount of desperate and pathetic.

Daisy considered that this was a ruse, but was attracted to this stranger. "I could help you with the menu," she said with a smile. "I'd enjoy the company."

Sophie Jean's analysis of Daisy was, of course, right on. Daisy was always willing to help, and that had gotten Chops in the door—literally.

They took seats on opposite sides of a booth. After introductions and some small talk, Chops realized that the longer he sat there without telling the truth, the less likely it was she would be willing to join them.

He cleared his throat and spoke directly. "I have to apologize, Daisy. I purposely ran into you this evening, and I want you know why."

Daisy looked at Chops a little uncomfortably, but because of Chops' engaging and direct communications decided to hear him out. She smiled and cocked her head, so Chops took that as a signal to continue. "I work for an organization that is doing a very important service to humanity," he said. "We want you to help us save the world." Chops paused, and then continued by telling Daisy about Moto and his organization and its goal of saving the world from AI dominance. A few moments went by and Chops looked to Daisy for an answer.

"I've always wanted to save the world," she said with a sweet smile. Chops touched her hand and the unexpected trust that Daisy felt for Chops was confirmed.

When the food came, Daisy ate in silence, without looking up, but afterward they took a long walk to discuss the logistics of reporting and how she would update the AI program with Moto's protective code. Daisy's arthritis began to bother her and they parted ways, agreeing to talk again soon.

When Daisy got home, she was bewildered about what had just happened—not so much at the secret organization to battle AI dominance but that she had so quickly trusted a complete stranger. There was something about his directness that made her feel at ease. She had just committed to be a corporate spy, but being a part of *this* team, with the noblest of causes, was exciting. She turned on the television in her bedroom and stared at it far into the night, unable to sleep.

Matter is energy. Moto had grown weary of continually updating his stealth code for operating system changes as it was overly time-consuming. During the past two weeks he had been in a manic phase, sleeping only a few hours a night, but had made a major breakthrough. Moto had long thought that computer hardware was inefficient, so the next logical progression was to develop a quantum virtual processor using only electrical energy - freed from the constraints of hardware.

The breakthrough came on day thirteen, when he launched a calculator that added 2+2 using only the electricity in the wall outlet, without the use of his laptop's processor. The highly sophisticated stealth technology he had developed at Ball and Chain was child's play compared to the possibilities of this new virtual computer.

Moto needed help developing this new technology, and Indy was his best programmer. Indy was a descendant of an aboriginal tribe in Australia, but he had inherited red hair somewhere along the way. In addition to being a top-notch systems programmer, Indy had some unique skills. He was seven feet tall, lanky, and looked a little clumsy, but he could run the hundred-yard dash in eleven seconds. He could also sing two tones at the same time, which made him sound like a didgeridoo. Moto was concerned about the talent drain at Ball and Chain, but he needed the most capable people with him fighting AI domination and Indy was key to the success of the virtual processor.

It was quite a sight when Chops walked into the safe room with Indy to discuss his role in developing Moto's quantum virtual processor since Indy was more than two feet taller than Chops. Being of aboriginal descent, Indy had a mystic ability to sense when something important was about to happen, and he could sense that Chops was about to manipulate him but for something crucial and life-changing. Chops sat down, and Indy felt that the world depended on him, feelings of vast importance, righteousness, and danger overwhelmed him.

Moto joined them and Indy's eerie feeling about the future continued, but the compelling technology and the excitement about his role in developing it helped him overcome his uneasiness.

Finishing up with Chops and Indy, Moto stopped by to chat with Hurley and Sophie Jean. Sophie Jean lamented that she did not have a large pool of trustworthy programmers to pull from to replace those Moto was taking from Ball and Chain, so for the long-term viability of Ball and Chain and the AI effort, she suggested setting up an in-house training program. If they could find people who could be trusted and had the aptitude, they could develop their own programmers from people without any prior programming experience. The curriculum, the teachers, and finding non-programmers with aptitude for programming could be done, but would take considerable time and effort. Sophie Jean stared blankly in Hurley's direction.

Hurley looked up and saw Sophie Jean's concern. "You know," he said with a serious but somehow comforting smile, "we can't cut corners when it comes to any of the work we are doing. It's all too important. Our search engine is about as sophisticated as it can get, so the time-consuming part is our decision-making and taking action. I think I can automate seventy-five percent of the decision-making and half of the rest, which would free up enough time in about six months to get started on the in-house training program."

Sophie Jean stared at Hurley, but now it was directly into his eyes. She considered how she was often just a rubber stamp for the action plan the computer recommended, but was still concerned about the control she would be giving up.

Internally, she reconciled herself to this course of action. "You're right, of course, Hurley," she said with conviction. She told Moto she would add some specifications to Hurley's programming, as anyone or anything that takes actions must be accountable and trustworthy, and she wanted to develop an audit program that assured just that. Moto understood Sophie Jean's concern that they were developing an AI solution to this issue and his stomach churned as he nodded in agreement.

***

In a room three doors down, Indy sat alone at a computer screen. Before him was the quantum programming language and application Moto had developed, and he stared with fascination at what Moto had done. The code was elegant and read like a prize winning novel.

Once the virtual processor was launched into the electrical grid, it had the capability to launch itself again. Moto hadn't stopped there: he had planned for media, such as water and gas. Indy picked up a graham cracker to munch on, his mind swirling with how to take Moto's accomplishment to the next level.

Chapter 8 – Lily and Moto Communicate

Moloana and Lily had literally been "down on the farm" for the past few weeks and most of the physical doomsday site set-up was complete.

It might be five weeks or five years before they would need to activate the doomsday bunkers, so they had to be ready. Samantha, a detail oriented master in camouflage, was chosen to manage the day to day operations until the sites had to be triggered. Samantha was once Moto's baby-sitter and they had spent endless hours playing hide-and-seek, Samantha often surprising Moto by appearing out of what seemed to him like thin air. Samantha had never married and had no children of her own, but she loved Moto like a son.

A mechanical engineer by trade, Samantha had lost some of her physical capabilities, but her knowledge and experience more than compensated. She knew every nut and bolt in every system in every doomsday site and relished the thought that her job was to keep everything in top working order without anyone knowing the sites existed.

The doomsday sites in Samantha's capable hands, Lily and Moloana spent most of their time developing AI countermeasures. Everyone had ideas for how to fight the threat, and Lily and Moloana acted on forty-seven of the sixty ideas received from the team.

Moto was surprised and encouraged by the new countermeasures submitted that he hadn't considered, and realized just how good his team was and how important teamwork was going to be in protecting humanity.

Lily was excited to hear about Moto's breakthrough with the virtual processor and literally dreamed about the ways she could launch apps into the electrical grid to combat runaway AI programs. In one dream sequence, she and Moto raced around inside the electrical grid, jumping and playing and running circles around the AI enemy programs as they laughed and chased each other. When she woke up from the dream, she expected the feeling to fade, but it didn't.

Later in the day, Moto texted Lily that he had dreamt about her the night before. They were walking in the woods when a thunderstorm came up, and they ran to avoid the lightening that was striking all around them. Moto didn't feel like they were in any danger and was enjoying the challenge of zig-zagging with her through the woods. Lily sent a text back that they needed to talk and that it had to be today.

Moto was just climbing out of his car when Lily arrived at his place that evening. She ran to Moto and gave him a long, tight hug.

Once inside, Lily quietly but firmly asked for a glass of wine, and Moto returned with two glasses and a bottle of a nice cab. There would be no small talk this evening. He lay on the chaise lounge while Lily climbed into her favorite chair and sat cross-legged, smiling.

"I liked that you dreamed about me last night," she began. "Do you remember any other details about your dream?"

"It was funny," Moto said. "We were running around the woods, avoiding the lightning strikes, and having a good time. If the lightning struck close to us, we would laugh and say, 'Ooh that was close.' I thought that was odd, because we should have been afraid."

"What were we wearing?" Lily asked.

Moto thought for a minute. "I think we were wearing all white. Wait, we both had sweaters on and they were black and white."

"Do you remember anything else?" Lily asked with a knowing look on her face.

"What is all of this about, Lily?"

She smiled and stretched her arms over her head, so Moto ran the dream through his head again. "There's one other thing I remember," he said, scratching his ear. "As we were running through the forest, we were looking for cover and saw some caves on the hillside— two rectangular ones above one round one. It almost looked like two eyes and a nose. The lightning struck around the two rectangular caves but not around the circular one."

"Did they look maybe like … electrical outlets?" Lily asked, knowingly.

Moto felt a shiver flow through him, knowing immediately that that is what they were. His eyes opened wide, he took a gulp of wine, and turned his head, looking deeply into Lily's eyes.

While Lily was listening to Moto describe his dream, she began remembering parts of her dream that she hadn't recalled before, and as she then recounted her dream to Moto, he remembered more of his dream. They talked back and forth about the details of their respective dreams.

"I've never remembered so much of a dream before," Moto said at last.

"Me neither."

Then together in unison: "What do you think it means?"

This set off a minute-long round of giggles that abated only when Moto choked out, "I hope this doesn't mean we're going to speak in one tongue from now on," which only set off another round of laughter.

"Let's talk to Sophie Jean in the morning," Lily gasped and, taking Moto's hand, she led him to the bedroom. "I think we're done here...." She said sweetly.

Chapter 9 – A Session with Sophie Jean

It was almost 9:30 in the morning before Moto and Lily could meet with Sophie Jean which gave Lily a little time to look up "parallel dreams", arming her with questions for their session. When they arrived, Sophie Jean started reporting on the in-house training program she was working on.

Holding up his hand, Moto interrupted her, "We have something different to talk to you about today," he said, with a big grin on his face. They sat in a small circle, each in a separate, deep, soft, white chair. Moto turned to Lily, who told Sophie Jean about her own dream, Moto's dream, and how similar and detailed the two dreams were.

Sophie Jean took off her glasses and set them on the table next to her. "I am going to give you a few possibilities of what the two of you just experienced," she began. "One is that this was just a coincidence," and she paused to see their reaction, but seeing their blank stare, she continued. "However, I must say that the probability of this being a coincidence is small due to the large number of common elements in your dreams, so it's probably not a coincidence. Another possibility is that, as you began to discuss and review your dreams together, false memories were inserted in both of your recollections, maximizing the perception that the dreams were similar and minimizing the differences." Moto sat back in his chair and Lily leaned forward, clearly waiting for the good stuff. Lowering her head, Sophie Jean said, "I take it you don't believe it was a coincidence or false memories."

Moto and Lily smiled and winked at each other and while Moto settled further into his comfy chair, Lily sat up straight, her legs

36

folded underneath. "Well," Sophie Jean said, "this leaves a couple of other options, neither of which has been proven to exist. An out-of-body experience is one that many believe is the releasing of the soul or identity from the physical body to travel elsewhere in the universe, allowing one to float through walls or ceilings or, as in this case, into the electrical grid. When people have such an experience, they often feel frozen or unable to open their eyes momentarily when they wake up before their consciousness returns to their physical bodies. Did either of you have that feeling?"

Moto and Lily looked at each other briefly before Lily said, "I felt just the opposite when I woke up. It didn't feel like I had been somewhere else."

Moto nodded. "I felt normal, like Lily and I had been spending time together."

Sophie Jean continued: "The lack of the sensation of coming back into your physical bodies does not preclude this from being an out-of-body experience, or OOBE, but it does open up another alternative. Scientists have identified a sort of neurological Wi-Fi that allows people to communicate feelings without use of the five senses. Usually, it results in a gut feel or group excitement like that at a football game. It is possible that you are connecting on that level. What are your thoughts?"

The mental Wi-Fi feels right to me," Lily offered. "I'd like to try to make it happen again." Moto smiled and nodded enthusiastically.

Sophie Jean laughed. "Okay then, here's a series of questions for you about what you were eating, drinking, and doing in the days leading up to this experience."

When they had finished answering, Moto and Lily walked out of the room, nudging each other and smiling. They did not notice Hurley, who had raised his hand to say hello, as they were too engrossed in each other to notice.

Hurley and Sophie Jean met every day at this time—not that Hurley wanted to - but Sophie Jean wanted to ensure that she knew everything Hurley was doing in case he'd forgotten to mention something important.

Hurley had been modifying his background-check routines to automate decision-making about potential candidates. There was often conflicting information about a candidate and those chosen had to be squeaky clean. Finding the truth on the internet, let alone the dark web, was a challenge. Sometimes even a source of information needed background checking. Not many people met the original goal of 100% surety, so Hurley lowered his surety goal to 98% to get enough people on board. Chops made the final decisions based on interviews with the candidates, though they didn't know they were candidates until after Chops spoke with them.

The prime age for a computer programmer was between 16 and 42, the years when mental ability was optimum. However, at 16, people didn't have enough life experience to determine their trustworthiness, and after 40, their skills began to decline, so Sophie Jean and Hurley set the search parameters from age 22 to 35. Since the in-house training was for people with no programming experience, they sought loyal, caring perfectionists with sharp, logical minds.

Some of the background information came from the dark web, which Hurley had been mapping for years. It didn't happen often, but occasionally he found a candidate there among the pedophiles, deviants, and serial killers who had either been compromised or was compromising someone else. Unlike the regular internet, the dark web was constantly changing and evolving in an undocumented way, and it was expanding exponentially. To keep up, Hurley's dark web scanner had to evolve with the environment, to learn and change as the dark web expanded.

Hurley's life had been shaped by an incident that occurred when he was just a boy. One day, when he was six, he and his parents were out on their sailboat, and a large speedboat rammed them. His mother was killed instantly, and after wrapping Hurley in a life vest, his father tried to help her, but he became entangled in the sailboat's lines and drowned as well. Hurley was rescued, but he remained terrified of speedboats. Even seeing one in the distance gave him anxiety. Sophie Jean attributed Hurley's desire to help people and commit to good causes to his inability to help his parents on that terrible day.

Hurley sat across from Sophie Jean. "Anything I need to know today?" Sophie Jean asked nonchalantly.

Hurley shook his head but didn't get up. Sophie Jean cocked her head as if to ask why he was still there. Hurley jumped up with a shy look on his face. "See ya'," he said over his shoulder and he turned and walked out slowly.

Chapter 10 – The Electric Grid Network

"This is going to make our life so much easier," Otis said.

"Yeah," Lucy sighed. "The effort to keep up with all the operating system updates has us buried."

Word of Moto's breakthrough with the virtual processor was exciting for the whole team - they would have virtually unlimited processing power while remaining completely undetectable.

Lucy and Otis were developing AI detection scanner technology, similar to virus protection but would render AI infiltration harmless. Lucy developed protection for computers world-wide, while Otis managed a task force defining launch points for the virtual processers. Ginger, Moloana, and Indy, who were on Otis' team scrapped the old internet tools and replaced them with virtual processors that ran on the electrical grid network.

Otis never expected to know this much about power grids, and he found it a little scary to see how much of the world's power generation capability used aging technology and how easy it was to hack. Indy and Otis added new virtual processors; their first app, which they called "Retriever," would load on the electrical grid at a power-generation site and travel on electrical current, identifying junction boxes, electrical outlets, and where devices were plugged in.

Ginger developed "Sniffer", an app that identified what type of device was plugged into an electrical outlet. Sniffer would ride the electrical current into the device and use a quantum electron scan to determine what the device was. She destroyed a couple hundred coffee makers while perfecting this technology. The app had limitations, as it couldn't scan the device unless it was turned on, but it could turn on most devices.

The virtual processor executed its tasks quickly without the constraints of a physical computer. When Retriever and Sniffer did their thing on the power grid, it took seven minutes to analyze fourteen billion devices. A two-percent drop in power capability during this processing was projected, so they slowed it down to just over two hours so that the power usage was undetectable.

Moto's quantum processor manipulated electrons in a pattern for storage, so that information on the 14 billion devices Sniffer detected could be stored on a copper wire only three inches long. Ginger tested storing data on external devices, including eight terabytes of data in the electronics of her refrigerator.

"Ha, pretty cool," Otis said when she told him.

"Funny," Ginger replied without looking up. Otis mumbled something about getting the cold shoulder and Ginger muffled a giggle.

Otis and the task force were struggling with accessing all the computers worldwide because there were so many individual power grids. From power generation to transmission lines to substations, the sizes of the grids varied widely. Particularly challenging were the smaller grids—solar farms that powered businesses and neighborhoods. Consolidating all of this information wasn't feasible, so they had thousands of individual data bases that weren't tied together. Otis asked Moto to come to a brainstorming session to help solve this problem.

"I apologize for withholding something from you," Moto began, looking around the Safe Room at his team. "I have a good and old friend named Malia. We used to play beach volleyball together back in the day." Moto paused a moment, a reflective smile spreading across his face. Then he said, "Come on out, Malia."

Malia emerged from behind an invisible panel in the back of the room. She was older than most of the team members, and although her gait was a bit slow, her voice was that of an eighteen-year-old.

"Moto asked me to talk to you about what I have been working on for him," she said sweetly, "but first a little background." She looked around the room, her eyes soft and sparkly. "When I was little, I always wanted to be invisible—it was an obsession with me—so I decided that, in my lifetime, I would figure out how to do that. Although I haven't been able to make myself invisible, along the way I learned a great deal about light and light spectrums and have done a fair amount of tinkering." She put on a pair of glasses which made her look like a professor, sat down, and folded her hands in front of her. "As you know, above the visual light spectrum are ultraviolet light, x-rays, and gamma rays, and below visual light are infrared, microwaves, and radio waves. As you move up the light spectrum, the wave lengths become shorter, so they move through solid materials more easily." Malia fixed her eyes on the table and continued, her hands gesturing in an effort to describe the indescribable. "As I worked to make myself invisible, I had two options: I could try to bend visible light around me so it didn't reflect, or I could convert visible light to gamma rays that would go right through me, reversing the process on the other side, effectively passing back the image of what was directly behind me. Unfortunately, neither method worked because, for one thing, it's pretty dangerous to have gamma rays going through flesh." Malia paused and smiled. "So now my theory is to use gravitons to make either method work."

The others in the room shifted in their seats and looked down at their hands, at each other—anywhere but at Malia. Malia cleared her throat. "Well, you're probably wondering where I am going with this. That's reasonable enough." Malia smiled, and the room smiled back, respectfully. "I am not here to tell you I have a graviton-based communication system that allows people to be invisible," Malia said with a laugh, "but I have stumbled onto something that can help you. I call it FRISBEE for Far-Reaching Infrared Symbiotic Bio-friendly External Electricity. It is a type of smart wave that detects what is in its path and uses the safest frequency to get through it while retaining ninety percent of its energy."

Jaws dropped, and Otis squirmed in his seat, bursting with a thousand questions.

Moto stepped to the front of the room. "Malia has agreed to give us this technology, which could be the key to our success connecting the electrical grids. She won't be joining the team, but I'll fund her invisibility research for as long as she wants."

Malia bowed graciously and said tongue in cheek, "I'll be available for questions and will see you soon—but you may not see me." She was out the door before she could hear the chatter and excitement that followed.

Robert Baron started the Minnesota State Bank of Dollars when he was nineteen years old. In the intervening fifteen years, he had grown the bank to a value of $175 million, most of it coming from properties he had repossessed.

Each property that he had loaned to was valued at over $2 million, and he had hired just one loan officer to oversee the thirty or forty loans in his portfolio. The wealthy were attracted to his bank because of the low interest rates for high-value properties.

It was the time of year when Robert Baron and his loan officer met to determine which properties to foreclose on. "Tom," Baron began, "we need to add $14 million to our bottom line this year. Which properties should we go after?"

Tom Fulari had worked at the Bank of Dollars, otherwise known as "BOD," for eight years, so he knew the drill. He slid a piece of paper across the table to Baron that contained the names of three people, owners of properties that were valued over $5 million. "First on the list," Fulari began, "has just divorced and is left with the house and a big alimony payment. If we run the stock scam and get him fired, he'll be penniless except for his 401K, which he won't touch. He'll file for bankruptcy to stop the alimony payments, and we can grab the house." Baron nodded in agreement.

"Next is the CEO of that chicken company that uses religion to deny his employees rights to healthcare and keep gays out of his stores. The funny thing is, he is gay himself, so we can blackmail him and take his house in exchange for letting him keep his $4-

million-a-year job." Baron grinned, a drop of saliva appearing at the corner of his mouth.

"Last is a $300 million lottery winner who is spending money like crazy. We could, you know, 'help him' "—Fulari used finger quotes here—"with his investments and pick up his $6 million house for a song."

Baron struck a match and burnt the piece of paper Fulari had given him. *This has to grow faster*, he thought; *I need to hire another loan officer*.

Robert Baron would have been the perfect candidate for his own foreclosure scheme. His personal spending had grown out of control. He had no friends, so he had to pay people for entertainment and companionship, but the arrangement was perfect for a narcissist: control without emotional ties.

Baron scanned a resume from a young man who lived in Altoona, Iowa—a resume that Riffy had carefully prepared to target BOD. There was ruthlessness in it that appealed to Robert, including experience in foreclosing expensive properties. Andy Inch's resume indicated he was a paraplegic who worked from home with top computer skills and knowledge of the dark web. Robert Baron picked up the phone to schedule an interview.

***

Riffy asked Edie to watch Andy Inch's video interview so she could improve her interface. Edie was surprised at the naturalness and timing of the AI program when responding to Robert's questions. The AI program had to read the facial responses of the interviewer to determine its own best answers.

Sometimes the Andy Inch program had no "best answer" to a question, so it allowed Riffy to choose among two or three possible responses. The program provided a delay by pretending to take a drink of water, cough, or sneeze so that Riffy had time to provide a response, but if there was no override from Riffy, the program chose the answer with the highest probability of success. As the interview progressed, the Andy Inch program became increasingly confident in its responses as it learned to read Robert's reactions.

Edie had wriggled her way into Riffy's confidence by going above and beyond, providing extra subroutines for the AI brain module. Yet, could there ever be trust between them? How could there be? She was a spy and saboteur, and trust was not in Riffy's nature.

***

Sophie Jean read between the lines in Edie's latest status report, seeing that Edie was beginning to respect Riffy. She asked Hurley to run a more detailed background check on Riffy so that she could re-analyze his personality. Once done, she arranged for a meeting with Chops, Hurley, and Edie to discuss the results. They all needed to be in agreement with what she was proposing.

They met in a little restaurant in the small town of Dallas Center, Iowa – not too far from Altoona. It was a rustic little hole-in-the-wall with large wooden tables and leather back chairs and barstools.

Once they were seated in a dark corner booth, Hurley quietly began, "The background check on Riffy brought up a felony assault charge in Houston for which he spent 100 days in jail, which automatically put him in the not-trustworthy category. But it turns out that Riffy was just rough-housing with a friend and it got a little intense. Riffy's friend's mother tried to intervene, and Riffy accidentally hit her, knocking out two teeth. She filed charges, and his friend lied to police because the complainant was his mother, so Riffy pled guilty to protect his friend. Later communications from Riffy's friend made clear that it had been an accident."

Sophie Jean continued, "We ran an analysis on Riffy's profile. Because of the felony conviction, Riffy was denied jobs that could have led to a great career. From what I see, part of his motivation is to make money, but he is strongly motivated to reward loyalty and fight for people who have been wronged.

There followed an exceedingly long pause considering Edie's proclivity for filling silences, but she finally burst out with "I knew it! He's one of us!"

Thanks for your work on this," Chops said, nodding at Hurley and Sophie Jean, before turning to Edie. "Normally I would speak with Riffy to recruit him for our team, but in this case I'd like you to talk to him."

"But what would I say?" Edie asked helplessly, looking around the booth at her three teammates.

"Just tell him the truth—everything. And tell him I want to talk to him," Chops replied.

Edie left to drive back to Altoona and Chops, Sophie Jean, and Hurley headed back to the Des Moines airport. "Is this a chink

47

in our armor?" Chops said directly to Sophie Jean. "We put a good guy in the bad-guy category."

"We could go back and reevaluate anyone with a felony conviction," Hurley replied.

"No," Sophie Jean said, her voice firm. "I think our process is exactly what we want. It's not likely that anyone with a felony conviction is trustworthy enough to bring onto the team. We eliminated 7.2 million people and only one was eliminated in error—"

"That we know of ... so far," Hurley interjected.

"Right, but it would be worse to make a mistake the other way," Chops said. "I think we're good, and we've learned that Edie has great insight into people's character. It is something to keep in mind for the future."

***

Edie returned to Altoona well after dark, and Riffy was standing on the front porch like he did every night. Edie rolled down the window and called out, "Let's go for a ride." It was a statement, not a question.

Riffy shrugged and got in the car. Edie stared straight ahead, her hands gripping the wheel. "I don't know where to start," she said at last.

"Start at the end," Riffy said.

Edie took a breath and then blurted out, "Chops want to talk to you about joining our organization."

"Hmm," Riffy said, revealing no surprise. "Well, then, who is Chops and what is your organization?"

Edie looked at Riffy with apprehension, "Chops is who I file my status reports to, and the organization is a clandestine group associated with the company Ball and Chain." The rest was easy, so Edie spit it all out at once. "I was sent here to infiltrate your organization because you are developing AI programming. We are trying to protect humanity by ensuring that AI doesn't get out of control and take over the world." Edie took a deep breath.

"So that's why you put the failsafe code in Andy Inch?" Edie's jaw dropped. "Yes, I saw it," Riffy continued. "I didn't remove it. It actually seemed like a good idea and I was going to ask you for the access code before you left. But you're so good, I wanted to squeeze as much of your expertise out of you as I could while you were still here."

Edie was so relieved that she grabbed Riffy and kissed him full on the lips. "You're a good soul, Riffy" she said. It was the first time Edie had seen Riffy smile.

The next day, Chops met with Riffy. Although nothing changed, everything had changed: Moto had another ally that he could use when the time was right.

Moto curled up around a pillow on his deep, soft loveseat. The television was on, but he wasn't watching. Every once in a while he heard something—an indistinguishable sound from outside— that perked his ears, but he quickly lost interest. His mind swirled with thoughts about his team and what they were doing. He felt like he wasn't doing enough. He wanted to dig in and help Indy with the virtual processor code, but he had done the groundwork and Indy was more than capable. When the next technological advance was needed, he would be back where he was most comfortable, in the trenches.

Moto put his phone on "don't bother me" but Sophie Jean got through his defenses and texted that she'd booked a house on Maui for the team and that "We all leave on Saturday and will be there for nine days." Moto stiffened and thought *what would nine days of lost productivity do,* but continued reading the text, "All projects are ahead of schedule. Trust me, this will improve productivity." Moto smiled to himself, realizing Sophie Jean had anticipated his concern.

"K, thx" Moto texted back chuckling about his brevity.

Sophie Jean and Chops had been planning this retreat for two months as the team hadn't had a break for a year and a half. She attended to every last detail with the purpose of building a stronger bond between team members, while Chops ensured nothing critical needed to be done while they were gone.

Moto arrived at the Seattle airport on time only because Lily packed his bags and drove him there. They had dressed for the occasion: Moto wore bright white summer pants with a classy green and white loose-fitting Hawaiian shirt, while Lily walked along side in her tight black-and-white halter top and white short-shorts. Chops arrived in a wild Hawaiian shirt, oversized shorts and sandals, and Moloana was decked out in a ruffled top with gray shorts. Hurley sauntered in wearing his black shirt and black pants, with the Gingers flanking him wearing white flowing dresses. When Otis and Lucy got to the gate, they looked like they were headed to Manhattan, rather than Maui, as Otis had on a brown sports coat with a loosened tie and Lucy was wearing a nicely tailored black and silver jumpsuit.

Sophie Jean brought Riffy with her and introduced him to all the team members and although only Chops had met him before, everyone knew about him. Riffy looked a little uncomfortable and smiled shyly, but after a group hug, he felt a little more at home.

Moto glanced over at Sophie Jean with a questioning look, since he knew there was one more person coming, but didn't know who it was. She tossed her head and pointed down the hallway with her eyes at an elderly man walking towards them. Moto followed her eyes, recognized the man, and yelled, "Grampa!"

At eighty-two, Eli was wiry and tanned and had long, white hair. His pace was brisk and confident, and it took him no time at all to cross the distance to Moto and give him a warm hug. When he smiled, he lit up the room. Moto had the same build and long black hair, and there was a piece of Eli's smile in his. Eli waved to Sophie Jean and announced broadly, "I'll take it from here, Sophie Jean!"

He introduced himself individually to each team member, explaining that he would be directing the activities on Maui.

51

Moto felt like a six-year-old again, but it was a good feeling. Eli put his arm around Moto's shoulder as they boarded the plane, letting Moto take the window seat just like when he was a kid, and Moto looked up at Eli just like when he was a kid.

As the plane started down the runway, Moto began breathing faster and his hands shook nervously. Eli noticed the fear in Moto's eyes and when the plane was finally in the air, he asked Moto how his company, Ball and Chain, was doing. "We've been very successful," Moto replied. "Didn't Sophie Jean tell you about what we're doing?" Eli nodded, and Moto looked down at his hands now calmed, saying "It's great to see you again. I've been so busy—"

Eli cut him off. "No worries, Moto. We have our own lives to lead. And you were always–how shall I say it–driven? I'm proud of you for how successful you've been and for how you're already giving back to society. I haven't done anything but play for the past few years." For the next several hours, Eli listened while Moto told him everything.

Sophie Jean had assigned the seats in order to enhance team building by having people sit together that had things in common, but didn't know each other that well.

Hurley and Riffy were both usually quiet, but before the plane took off, Hurley broke the ice. "I know all about your past, you know, Riffy," he began, "and I have a lot of respect for what you did. Not many people would take a fall for a friend like that."

Riffy had been the bad guy for so long, he didn't know how to react to being respected. "Thanks, Hurley," he whispered. "That means a lot." To their surprise, they opened up as the flight went on, telling each other their thoughts and fears. Eight times one of them said, "I've never told anyone this, but...." When the plane landed in Maui, they were the best of friends and laughing at everything. It had been a long time since either of them had felt they had a trusted friend with so much in common.

***

Moloana had always felt a little uncomfortable with Ginger's focus and determination at work and that Ginger seemed unfriendly. When Moloana got to her assigned seat on the plane, Ginger was sitting by the window, but turned to Moloana and asked sweetly, "Would you like the window seat?" Moloana nodded, surprised that Ginger was so gracious.

As the plane climbed, Ginger told Moloana how great the trip was going to be and how she was looking forward to forgetting about work for a while. "When I'm at work, I have to focus so hard," she said. "I wish I could be more like you and be friendly at work, but I have to really concentrate or I get distracted." This was a side of Ginger that Moloana hadn't understood. She had always thought Ginger was just unfriendly—cold even. They talked about everything–Ginger's twin sister, Moloana's long hours and non-existent social life, and how they both played the timpani. When their conversation finally paused, they both fell quickly and comfortably asleep, waking up only when the wheels touched down in Kahului.

***

Otis expected to sit with Lucy when he got on the plane, but Lily was already in the seat next to his. Otis thought Lily was beautiful, but then he was attracted to everyone on the team, men and women.

One day, Otis and Moto were leaning against a wall, talking, and Otis turned and kissed Moto. Moto kissed him back lightly but then just turned away and continued the conversation. From Moto's point of view, it was like getting a pat on the head, as he was an affectionate person, but Otis thought it had been more than that.

Quite suddenly, Lily turned to Otis and said, "So, do you think Moto is a good kisser?" Otis' eyes opened wide–he wished he didn't have that tell when he was surprised—but before he could respond, Lily smiled and said, "Moto just mentioned that you kissed him one day. Did you ever really tell him how you feel?"

"No, not really," Otis squeezed out uncomfortably.

"Well, how *do* you feel about Moto?"

Otis had never really thought about it—he had only felt the attraction—but he was attracted to everyone on the team, so it might not mean much. Lucy was different, though: with her he felt a responsibility, a bond.

"I love Lucy," he blurted out, a few decibels louder than he had intended. They looked at each other and began to giggle about his inadvertent reference to the 1950's TV sit-com. In the next moment, unexpected tears ran down Otis' cheek. Lily reached over and held his hand, and they slept that way the rest of the flight.

Lucy sat down next to Ginger. "You know, I can't tell you and your sister apart. When I see one of you in the hall, I just say hi, Ginger," Lucy said bluntly.

"You know, Lucy," Ginger said, "we figured you always thought it was me because you pronounced Ginger like my name is pronounced, not my sister's."

"Oh, uh, I, uh," Lucy stammered, "Are your names pronounced differently?"

"Yes. My name is pronounced Ginger, while my sister's name is pronounced Ginger," she continued with a straight face as Lucy's face reddened, "and we really don't look all that much alike. Most people can easily tell us apart."

Lucy looked around to see if another seat was available.

Ginger laughed. "I'm sorry, Lucy," she said with a warm smile and a hand on Lucy's arm. "I was just teasing. Nobody can tell us apart and practically no one can pronounce our names right."

Lucy's eyebrows came down into a V, nearly obscuring her eyes. "Whatever," she growled, crossing her arms.

Ginger knew all too well about being angry and having to control that anger. Until she turned twenty-two, she had been angry all the time—with her sister, her parents, her friends, even strangers who had the misfortune to cross her. She lashed out at everyone. After spending eight months in anger management therapy and learning the warning signs for the onslaught of that unpleasant emotion, she made a few friends, and the relationship with her sister improved. She learned she had been frustrated that she wasn't a unique individual and resented everything about her sister—her name, her looks, her everything. She learned to work on developing her own personality as an individual, and her resentment faded.

She saw in Lucy the same anger problem she had faced. "Hey, Lucy," she said brightly, "did you know that, until I was twenty-two, I was pretty much a tomboy?"

"Yeah," Lucy replied, "me, too."

"It was more of a rebellion and fear of failure than wanting to be one of the guys. I was angry and frustrated most of the time."

Lucy's shoulders relaxed a little. "What changed? You aren't like that now."

"I went into anger management therapy and I learned why I was so upset all the time and how to control it. I had been acting like a tomboy so I wouldn't fit in - an excuse for failing to be part of the world. My anger was a circular thing: I withdrew because I was angry, but then I was angry because I had withdrawn."

Lucy was quiet for a few minutes. Finally, in the tiniest voice possible, she said, "thank you, did you ever build a treehouse?"

The rest of the conversation would be the first step in a powerful friendship between two of the feistiest members on the team.

***

Sophie Jean hoped that her grand seating plan would bring the team closer together during the long, dull flight. She and Chops sat together. "Are the kids behaving themselves, Ma?" Chops asked wryly.

"I reckon so, Pa," Sophie Jean leaned over with a smile, "I saw Hurley laughing." Chops shrugged with surprise.

They arrived in Kahului a little early but had to wait until the last of Eli's four suitcases finally tumbled down the chute into baggage claim. A stretch limo awaited them, and half an hour later, they pulled up to a huge twelve-bedroom house directly on Keawakapu Beach in the lush suburb of Wailea in South Maui.

Once settled, they wandered out to an expansive back yard lanai and a view of the setting sun over the Pacific. It was 73 degrees, and a light wind caressed the surface of the infinity pool, which was surrounded by lounge chairs and tables. A smiling server took orders from a small menu and placed "pupu" platters at each table.

"Welcome to paradise!" Eli boomed, raising a frosty glass of lemonade, as he had forbidden alcohol. "For the next few hours, do what you want, but at Maui midnight, which is 9:00 PM sharp, be back in your rooms with lights out!"

As Sophie Jean had asked him to do, Eli led the team by organizing the next nine days of activities in ways that would form relationships, recharge minds, and build a more cohesive team. He had them playing Frisbee and paddleball on the beach, hiking, zip-lining, swimming, snorkeling, and paddle-boarding. They played (bad) golf one day and tried their hand at pickle ball the next. They ate the local food. Hurley called it boot camp in paradise.

The morning that they had to pack for their flight home came too quickly. They gathered to generate new ideas and suggestions, just as Sophie Jean had hoped, before climbing into the waiting limo.

Moto gazed appreciatively at Sophie Jean, as she had accomplished what he hadn't known needed to be done.

Waving and shouting aloha, Eli watched the limo cruise out of sight, staying behind to take a few days for himself.

Moto was relieved that he had managed to fend off his manic state until he got home. The laid-back atmosphere of paradise had almost driven him over the edge, but now he could just let his mania take over.

Hunched at his desk, he poured over an article about the development of a new type of photon—one that was combined with matter—that retained the memory of the matter as it passed through space. His mind raced as he thought about the discovery's possible implications for Malia's work on invisibility and for improving the method they were using to connect the electrical grids. He glanced at the clock—2:00 a.m. He didn't care. He called Malia.

Malia recognized Moto's voice, but couldn't quite grasp what he was telling her. Suddenly, Moto heard her clacking on her keyboard. "Give me a minute, Moto!" Malia barked, trying to concentrate. After a few minutes, she was up to speed and they talked for hours about this new discovery's applications. At the end of the conversation, Malia agreed to split her time between her invisibility research and helping Otis and Indy with how this new discovery might improve the ability to communicate between electrical grids.

Next, he called Sophie Jean, asking for her thoughts on the success of the Maui trip. Sophie Jean provided a summary, but asked that they review the details when she was "awake".

When Moto went manic, he was driven to write programs - something – anything - so the team knew to fill his inbox with specs for the most difficult and time-consuming programs.

Moto settled into his couch with his laptop, and two days later, his inbox was cleared out, the coding complete, and he slept for twenty hours, his laptop still open on his lap.

When Moto woke up, he had an uneasy feeling as the first thing he saw was an email from Eli:

> My dear Moto,
>
> I am so proud of you. It was wonderful to be a part of what you're doing for a short while and to see you one last time. I had a wonderful life but the body has given out. I love you.
>
> Grandpa Eli

Moto was shocked and quickly called Sophie Jean. Speaking in a whisper, Sophie Jean told him that Eli had had pancreatic cancer and only a few weeks to live. He had wanted to be useful one last time and spend that time with his grandson, requesting that Sophie Jean not disclose his secret.

They talked until Lily showed up at Moto's apartment. She listened quietly and comforted him as Moto let his emotions flow.

***

Eli began executing his plan right after the team left Maui. He mixed two bottles of pain killers with Irish whiskey and Irish cream in a flask, sent his time-delayed email to Moto, and smoked some weed to ease the pain. Once it was dark, he carried an inflatable paddleboard down to the water, along with some weights and a backpack that contained all the information the authorities would need when they found him. He took off his clothes and hung them on the bushes near the beach. And as he pushed the paddleboard into the surf, he felt relieved.

The water was calm. He paddled for hours, passing Molokini. The stars were bright, and the moon was almost full.

He sat on the paddleboard and put the weight belts on his legs and torso. Then he tied the backpack to the paddleboard and lay on his back, his head on the backpack. He opened his flask and took a sip. He looked at the sky, sipping from the flask, and sighed at the beauty surrounding him.

He fell asleep and rolled off the paddleboard sinking gently toward the bottom of the ocean. His last memory was that of the broad, grateful, loving smile Moto had given him as he left.

Chapter 14 – Beethoven Comes Home

The Gingers got reports from twenty-four development sites in addition to those Daisy, Beethoven, Edie, and Remi had infiltrated. After eight months in Ammons, Idaho, Beethoven was called back, as the Gingers determined that the virtual programmer that Londo was trying to build was never going to be a threat; the design and code were simply not sophisticated enough to work in a real-world environment. Beethoven had inserted a "dumb" switch into Londo's AI program that would cause it to go into infinite loops and delays at random times, effectively making it appear slow and unable to make a decision.

Beethoven looked forward to seeing his friends. He hoped his new assignment would be more satisfying, as Londo had directed him to write code that he knew wouldn't work, and then blamed him for the failures.

Ginger met him at the airport and went through a detailed debriefing, confirming that Ammons, Idaho, was no longer a threat.

"You did a great job, Beethoven," Ginger said when they got into the car. "We don't have another undercover assignment for you, but we're going to assign you to work with Otis and Indy. You'll love it. It's challenging and deals with new technology."

Beethoven looked down at his hands.

Ginger laughed and patted his shoulder. "I won't go into detail right now - Otis will explain it all. Trust me, you're gonna love it!"

Beethoven straightened his shoulders and said in the heartiest voice he could manage, "Thanks!"

Two weeks after Beethoven returned home, Londo was hit by a car crossing the street near his home. No one had seen the accident, and Londo had crawled three blocks trying to get back, only to die a few feet from the front door.

* * *

Currently, of the twenty-eight (now twenty-seven) AI sites Ginger was monitoring, only one was infiltrated by a staff member from Ball and Chain. That was Remi in Assisi, Italy. Remi had been in Italy for nine months, and her status reports were sometimes several days late because she was under tight control. She hadn't yet been able to install the failsafe code in the AI programs, and Chops and Hurley hadn't been able to finalize her extraction plan. Hurley determined that the Mafia was working with the Cartel on this effort, and Remi's latest reports suggested links with the Russian mob and the North Korean government—possibly others—but she couldn't get close enough to get a handle on the full scope of the organization.

Three days later after Moto had read Eli's email, Chops called a meeting in the safe room regarding Assisi, Italy. Riffy, now a part of the core team, flew in for the meeting, and when Moto finally joined them, the room quieted.

"Hi, team," Moto began, "thanks for all your kind thoughts about my grandfather." He paused and looked around the room, seeing the nods and silent well-wishes, his eyes stinging. Then he cleared his throat and pushed on. "We're coming up against our biggest challenge so far, so I need everyone to focus on how we should go forward." He gestured to Chops, who hopped down from his chair.

"Listen up," Chops began, "when Moto started this effort, we all had our own expectations about what we would come up against. But now we've discovered something very big and very dangerous."

Ginger took a sharp breath, as the danger Remi faced felt even more real to her.

Chops glanced at her and leaned forward with his hands on the table, his eyes seeming to make contact with everyone in the room. "One of Ginger and Ginger's operatives, Remi, was assigned to work on an AI development project in Cozumel that was funded by the Mexican drug cartel. When Remi was deployed, they immediately sent her to Italy, where she has been working ever since. We confirmed that the Mexican drug cartel is working with the Italian Mafia and recently determined that the Russian mob and the North Korean government and maybe others are also involved. We don't know the extent of what they're developing, but we believe the Italians are the lead organization, at least for now."

Chops heard the sound of people squirming in their chairs and whispering to each other. The "not knowns" in what Chops had said were plentiful.

He looked down at the table and then continued, "This organization, if it is as large as we fear, has ten times the funding we have. Remi found that her small group is isolated from the other groups, which aren't on the Italian power grid but on individual generators and solar power systems. They don't communicate with the other teams at all. The programs are offloaded to external drives and hand-carried to the other sites. We've told Remi not to ask any questions, as it would be too dangerous. She has infiltrated an international crime organization that, without her, would still be hidden from us." Chops took a long breath. "So where we are now is that we don't know what they are doing, who they are, where they are, or how they tie it all together. We need a plan."

The team was not a wimpy group of scared programmers; it was a focused group of dedicated professionals that could not be distracted from reaching a common goal. Otis broke the silence: "Indy is developing an inter-electrical grid communication technology that should be prototyped in seven weeks and that could help us find these other sites if we could get our receptor code into Remi's hands."

Ginger stiffened and said, "We'll need a comprehensive plan to protect Remi."

"You're right, we must protect Remi," Chops said firmly.

The team continued the discussion, but Riffy remained silent until the team came to the consensus that they needed more infiltration than Remi could provide. It was only then that Riffy spoke up. "I'll do it," he said, his voice cracking a little. Riffy regained his composure and continued "I can infiltrate the cartel-mafia-mob-Korean-whatever organization. I'm the only one with a felony conviction and the connections in prison to be able to do it."

Hurley looked at Riffy, his new best friend, and nodded his head. There was an awkward silence. "Hey, I'll be back," Riffy tried to say convincingly, "I trust you guys to make it work." But there was a significant chance he wouldn't make it back.

Chops continued the discussion on the technical issues involved with the infiltration. The team named the group they were infiltrating RICKY for "Russian Italian Cartel Korea Yada-yada-yada." Within a few hours everyone had their initial assignments and Chops went home to spend the night developing the plan. Otis helped him until every last detail was in place.

Sophie Jean went home to consider the psychological impact that this new conflict would have on the team. This enemy was dangerous, and the team's execution had to be flawless. By late into the night she had identified the personnel risks and fired the document off to Chops. Having done all she could, she fell into a deep sleep.

Remi was not completely aware of the danger she was in, but she did know that she had two masters–her Italian bosses and Ginger—and she toed the line with both. She received a communication from Ginger to initiate Strategic Alternative Nuanced Deception, or SAND, at 9:00 the next morning. Surely something important was about to happen.

Before she left her room for work, she ducked into the back corner of her closet and cut open the lining of her backpack. She pulled out the pill that Ginger had told her to take, slipped it into her pocket, and walked the 100 yards to begin her workday. At 9AM, she placed the pill under her tongue and let it melt.

The next thing she knew, she was lying in a hospital bed looking up at Ginger, who was dressed in a nurse's uniform.

"Non provare a parlare. Sei in ospedale," Ginger said broadly. Then she asked, "Hai bisogno di andare in bagno?"

"Si" Remi said dutifully and nodded.

Ginger helped Remi into a wheelchair and pushed her across the hall to the bathroom. Once inside, she gave Remi a long hug and then, her voice low and urgent, said, "We only have a few minutes. This room isn't bugged, but your room is." Ginger began detailing what Remi was to do.

"I can get a lot of the information we need if I get closer to my supervisor," Remi said during a pause. "He has a thing for me."

"It might come to that, but it has to be Plan B," and Ginger went over everything again, Remi repeating the plan back to her.

"Remember, Remi," Ginger said, her hand firmly on Remi's shoulder, "everywhere you go, your conversations are overheard, so be careful."

Suddenly, Riffy opened the door to the bathroom and looked directly at Remi. "Scusami" he said loudly with a wink and closed the door quickly.

"That was Riffy" Ginger said quietly. "He'll be back here in a week or two".

When Remi went back to work the next day, her bosses accepted her diagnosis of acute hypertension, but they did not know that she had come back with computer code that would allow the team to access RICKY's "off the grid" network.

The next morning, Remi dropped an eyeliner pencil in the corner of her bedroom, near an electrical outlet. She knelt down to pick it up, and hiding from the security cameras, put the end of the pencil into the plug, leaving a tiny microprocessor that installed a virtual endpoint into the electrical grid. Remi sighed with relief as she removed the device, stood up straight, walked to the bathroom and used the pencil to accent her large round eyes. *I can do this*, she thought.

***

Riffy returned to Altoona for the next phase of the bank project. Andy Inch, Riffy's AI program, had received an offer from the bank manager, and Riffy had left instructions for Edie on how to handle it. His main objective was to ensure that Andy Inch had exclusive control of the properties in his portfolio after three months of employment. The final video conference between Andy Inch and Robert Baron was scheduled for that afternoon.

Riffy walked into the office in his usual casual, confident way, and Edie screamed in surprise. She covered by yelling obscenities about her program code not working, then apologized to everyone for her outburst and went back to work as if she hadn't noticed Riffy was back.

The other programmers took turns going into Riffy's office to give him updates, but when Edie finally went in, she closed the door and collapsed into Riffy's arms, sobbing, overwhelmed. She caressed his neck and held him for a few seconds before Riffy gently pulled away. "We have a lot of work to do before the interview at 3:00," he said, looking warmly into her eyes. Edie nodded. She had missed him so much.

The videoconference began right on time. As the interview progressed, the sticking point was the exclusive control of the loans that Riffy required, but Riffy got more than he expected by playing to Baron's greed. Before long, Riffy's creation was hired with all the conditions he needed.

Later that afternoon, Edie picked up a scruffy-looking hitchhiker, who was walking down the side of the highway, dressed in dirty, ratty clothes. She and Riffy drove to their favorite rendezvous, a dimly lit sports bar with plenty of dark corners. They sat in the back of the room, away from other tables.

Edie was normally the center of attention, but she had to be the listener tonight. "Tell me about Maui," she began.

Riffy cracked a peanut shell and tossed it on the floor. "It was really fulfilling," he said. "I made a great friend in Hurley, and they all accepted me as a part of the team. But it's all because of you – you believed in me first." Riffy put his hand on Edie's and ran his fingernails over her skin, accidentally scratching her. Edie slapped his hand away, laughing. "See?" she said. "That's what happens to me when I try to be nice to someone—I get whacked."

Edie caressed Riffy's neck and kissed him. Riffy looked dreamily into Edie's eyes "Maybe this is a good time to tell you the rest," he said, his eyes turning focused and serious. "I want to talk to you about what happened in the safe room."

"I know about the safe room but I've never been there."

"Do you know Remi?" Riffy asked. Edie nodded. "Well, Remi was assigned to infiltrate the cartel in Mexico that was developing AI. Come to find out, though, It's not just the cartel— it's the Mafia and a lot of other really nasty organizations that have gotten together and because of my background—"

Edie leapt to her feet, scattering the peanut shells on the table. "No!" she yelped.

Riffy gently pulled her back to her seat, but there were tears in her eyes. They looked at each other for a long time before Edie recovered and Riffy finished telling her the plan.

"Darts!" Edie shouted, grabbing Riffy's arm, determined to enjoy their time together. "Let's have some fun!" Riffy smiled as she dragged him across the room to the dart board.

***

The next morning, Edie softly kissed Riffy goodbye and went home to get ready for work. As the door closed, Riffy rubbed his eyes. He had never been so happy. It felt good, but he figured it probably couldn't last—it never had—but this time it might not last because he thought he might not survive his assignment.

As Riffy headed to the office, he contemplated the busy day ahead. They had to finalize the parameters for Andy Inch's employment and do additional testing on some of the modules that had recently been enhanced. In a few days, the AI program would have a job of its own – loan officer. They would get activity reports to monitor the effectiveness, but on Monday, Riffy's creation would be Robert Baron's newest employee.

Developing an AI program that could perform a human being's job exceeded Riffy's early expectations, but Edie's expertise had been the key. They made a great team, and she was the glue that held that project together. But Riffy was the designer, and his insight into how to make this AI "being" come to life was so sophisticated that even Riffy didn't fully understand the extent of his own brilliance.

When Riffy sauntered into the office, Edie looked up briefly and winked at him. Riffy didn't flinch, but it took a certain amount of self-control not to grab and hug her.

Riffy asked the staff if there were any outstanding issues and thanked them for their hard work, since their employment was ending. He announced a party later in the afternoon. They were all well paid for their efforts, knew the project was ending, but did not expect any additional thank you, but they looked excited at the prospect of a celebration.

The party started at 4:00. Most of Riffy's employees liked seafood, so shrimp, catfish, mahi mahi tacos, and parrot fish were on the menu. Edie loved bacon and ribeye, so that was there too, but she was allergic to chicken so that was left off. The music began with Cat Stevens, then a little Snoop Dog.

Riffy made a short thank-you speech and promised everyone a cut of the profits, which would be considerable, including a $100,000 surprise bonus on this, their last day on the job - this loosened up the gathering quite a bit.

After Riffy sent them on their way, bonus checks in hand, Edie and Riffy sat down at a table holding hands and sipping wine. In a few days, Riffy would go to his new assignment, leaving Edie to monitor Andy Inch. Riffy's success in Italy would depend on the AI program's success as a loan officer – not only Riffy's success, but his survival was at stake.

Chapter 16 – Moto Has a Big Day

Moto and Lily had been performing the dream-preparation exercises that Sophie Jean had given them for several months. She asked each of them to think of a phrase before going to sleep and to associate that phrase with something they had in common. The next morning, they would see if they had picked up the other's phrase in a dream. The exercise usually failed because they couldn't remember their dreams or their sleep schedules were out of sync. They documented what, when, and how much they ate during the day, their physical activity, and their mental state, but hadn't made progress toward communicating again in their dreams.

Sophie Jean warned them that it could take years to reconnect in this way, but they were still frustrated. Secretly, Sophie Jean planned to change tactics after six months, as there was no point in continuing the exercise beyond that. But one Saturday morning, Lily called Moto, as usual, to discuss their dreams. "Mornin', Moto," she cooed. "How'd you sleep?"

Moto yawned and mumbled, "I'm at a loss for swords."

Lily gasped. "That's my phrase!" she exclaimed.

Moto wasn't surprised. He knew it had come from Lily. "And how did you sleep, Lily?"

"I can't plantain."

Moto was silent. Then he said, "Come on over. I bought some plantains for breakfast." They shrieked with joy.

"One other thing," Lily happened to mention, "I ate an orange right before I went to bed."

"I had an apple," Moto said quickly before they hung up.

Lily threw on some shorts and a tank top and jumped into the car. Moto met her with a huge smile. "Wow," was all he could manage when he saw her.

"Yeah, wow," Lily squealed.

After they were settled on the couch Moto said, "I invited Sophie Jean over. I figured she'd have some insights." Before Lily could reply, the front door flew open and Sophie Jean burst in.

"AAAAAAAHHHHHHH!" Sophie Jean yelled, grinning widely. She ran to them and hugged. "Nice job, guys! Tell me everything! "

They started in, Lily first and then Moto, then Lily, then Moto again until Sophie Jean stopped them. "I'm pretty sure I know why this worked" she said, "The trigger was things you have in common but are opposites!"

She didn't have to say anything more. "Of course," Lily said. "Even though we both like apples and oranges, apples and oranges are considered opposites."

Sophie Jean nodded. "Opposites attract, so your triggers need to be something you have in common, but in opposition."

Moto perked his ears and said, "Can we go to sleep now? I can't wait 'til tonight!" And they all laughed.

"I've been expecting this day for a while," Sophie Jean said, "but given the unknown nature of how it actually works, we were really lucky."

After reflecting on their success, Sophie Jean got up and stretched. "I'm headed to the beach today," she said. "Anybody interested?" Moto and Lily leapt up, all but circling each other in excitement. Sophie Jean had loaded her car with a cooler, beach chairs, umbrellas, and towels, and they spent the rest of the day at the beach, relaxing and basking in the glory of their success. It had been some time since Sophie Jean had had this much fun. After she dropped them back at Moto's place, she called her mom and made plans to spend the rest of the weekend with her.

Back at Moto's place, Lily looked closely at him for a few moments and then said a little sadly, "Let me know when you get back, Moto."

Moto had little control over when his manic state took over, though he could delay it for short periods. It was probably all the excitement of the day. "Sorry, Lily," he said, his eyes darting to his computer.

Lily forced a weak smile. "Get to work!" she ordered, "Everybody's counting on you!"

Moto spiraled into himself as Lily quietly closed the door behind her.

The next few days were the most productive of Moto's life. Lily came by every few hours to make sure he ate something, slept, and didn't lose touch, and he kept her up to date on what he was doing. He had completed the inter-electric grid communications code in cooperation with Indy, Otis, and Malia, in less than a week when they thought would take six months.

When Indy, sitting in his office, saw the completed system, he stood up and applauded. There was no one there to see him, it was a spontaneous reaction, and a tribute to Moto that Indy had no choice but to make. He took a selfie of himself cheering and giving the thumbs-up and sent it to Moto.

The technology was self-replicating and self-updating. The communication end points were virtual processors identified by their GPS locations, and they connected the electrical grids using the next nearest end point. Once the technology was deployed around the world, Indy developed a world map showing the status of the electrical grids, installing a series of monitors in the situation rooms of the doomsday sites and the main office.

Once Moto had recovered, he returned to the office and looked at the monitor. He nodded briefly, and then turned away. Moto had never been one to dwell on success or past accomplishments; he always looked forward. After he thanked Indy (briefly) for putting together the world view monitors, he brought Indy and Otis into his office to discuss the next project.

Moto shared his ideas about what to do with all the data they had, as they had access to every device in the world that was plugged in to an electrical outlet. As they brainstormed, they brought other team members in, one by one, until everyone was in Moto's little office, standing, leaning, or sitting on the floor. Chops was barely able to control the wildfire of ideas, but after several hours had prioritized the next several months of development projects.

Chapter 17 - Indy

Indy watched the monitor showing the progress connecting all the electrical grids. He couldn't believe how fast it went. He paged forward to the database display that counted the number of devices that had been accessed. With 6 billion people in the world, he wasn't surprised that there were 7 billion television sets and 8 billion computers, but 9 billion night lights? He paged through to find the number of coffee pots and snickered. *We could take over the world with coffee pots*, he thought.

"Hot dog!" Otis cried out when he came into the room. "Look at all the coffee pots!"

Indy smiled at Otis and then took a sip from his mug. "It couldn't have been done without you, Otis," he said sincerely.

Otis laughed, still looking at the giant display "Yeah, this is *grounds* for a celebration," Otis said still chuckling.

"No, really," Indy said, a hand on Otis' shoulder. "So many times, when I was going down the wrong path, you spoke up and convinced me to make the right decision."

Otis turned to Indy, only then realizing he was being complimented. Otis recalled several times, when he had convinced Indy to take a different direction, but hadn't felt that he'd had that much to do with the team's success, but with Indy's feedback he realized for the first time how integral he had been.

Indy continued, "I've kept Moto and Weetzie up to date on everything you have done to help me." Otis wasn't sure why

Weetzie needed to know what he was doing and stammered something unintelligible.

Indy pointed at the screen and cried, "Look at all the coffee pots!" and Otis smiled at the big guy in admiration.

***

Otis got to the beach bar a little early, but Lucy was already waiting for him in a booth in the back. When Otis sat down, Lucy put her hand lightly on his thigh and whispered seductively, "How was your day, Otis?"

As Otis told Lucy about his conversation with Indy, she moved her hand to his inner thigh and then higher, touching him. "I love you, Otis," she whispered.

Otis could barely speak, but he managed "I love you, too, Lucy," gulping, "you do know that I—"

"That you like men *and* women," Lucy finished his sentence with a wry smile "Well, I'm *one* of those two."

Otis smiled broadly, relieved that Lucy was so open-minded.

They sat in silence for a minute until Otis looked down at his lap - Lucy's hand was still there. He motioned with his head that they should leave and they got up before a waitress could take their order.

***

When Indy got home, a red light was flashing on his antique analog answering machine. He listened carefully to the message and became concerned about what he heard. He spent the night tossing and turning, considering what to do and when he awoke after only two hours of sleep, he had made his decision.

***

Everyone noticed as Indy loped into the office and headed to Sophie Jean's door, but before he could knock, she opened it and beckoned him in.

She sat back down in her chair and cocked her head. "So what's going on?" she asked. Indy hadn't spoken with anyone about his phone message or told Sophie Jean he was coming to see her, but he had a presence.

"I have to take a leave of absence." Indy said. "My grandmother in Perth is going on a pilgrimage and she needs me to come with her."

Indy's grandmother had left a message on his answering machine that said, "Indy, are you there," – the rest of the information had come to Indy in a dream.

"When do you leave?"

"Tonight."

Sophie Jean picked up the phone and called Otis and Chops, telling them to clear their schedules to work with Indy. As she hung up the phone, Moto burst in.

"Sophie Jean," he blurted out, "I had a dream that—" he stopped short and looked at Indy with wide eyes.

Sophie Jean finished his sentence: "That Indy was leaving?" Moto's mouth opened but no sound came out. "Don't worry— it's only temporary, a family thing," Sophie Jean continued.

Indy looked at Moto with his large, sad eyes and hoped that he wasn't upset about his leaving.

"Thanks for letting me know," Moto said with a sly smile, "even if it was in a dream." Moto locked into Indy's eyes briefly, confirming to Indy that he understood the message he had sent him in the dream, rubbed Indy on the back of his neck, exchanged glances with Sophie Jean, and left.

Chops strode in. "You know, Indy," he said, "this is probably the best time for this to happen, if it had to happen. You've just implemented the electrical grid communication project and we'll just be doing a lot of data analysis for the near future, which is Otis's strength. But when you come back, you'll still be our 'go to' guy."

Otis strolled in. "Hey, big boy," he boomed in Indy's direction, "we're gonna miss you."

"Oh, hi – what's your name again?" Indy said, with a deadpan expression and then laughing, stood up and grabbed Otis' hand.

Chops and Sophie Jean had long since developed replacement plans for everyone on the project, and when Weetzie walked into the room, Indy suddenly understood why he always had to copy her on his status reports. Weetzie, Otis, and Indy moved to a separate conference room and spent the rest of the day transitioning Indy's work. At 5:00, Indy was dragged out and subjected to everyone's well wishes before he hurriedly escaped for the airport.

"Any baggage, sir?" the agent asked.

Indy shook his head, "didn't have time," which cost Indy an extra fifteen minutes going through security, but he got to the gate with a few minutes to spare. Once on the plane, he tried to figure out what time it would be when he would arrive in Honolulu and then when he would get to Perth, but the time difference and international dateline were too much for his exhausted brain. He fell asleep and dreamed that his grandmother came to him with a knowing smile. It made him confidant that this next journey was the right thing to do.

The energy in Perth was electric when Indy and his grandmother embraced at the airport. At that same moment, in the office at Ball and Chain, everyone looked up, feeling a sense of excitement and warmth. Indy thought about the people he had left behind and smiled, comfortable knowing they would remain connected while he was gone.

Chapter 18 – Riffy Gets In

Riffy left for his assignment to infiltrate RICKY - the plan was for him to "run into" a couple of guys he spent time with in jail. They weren't really friends–Riffy had mostly kept to himself–but they had commiserated a few times. Hurley determined they were on the fringes of RICKY, and Sophie Jean had analyzed their personalities, determining that they were not particularly smart (no surprise for a couple of ex-cons) but wanted to impress their boss with something big so they could rise up the ladder. This was never going to happen, but the play was to get them to try.

Riffy took the train from Newark to Grand Central Station and checked into a posh hotel before heading to Milk and Fish, a dive his ex-con targets frequented. He didn't know how long it would take to run into them, but he was prepared for the long haul. He sat down at the bar and asked the bartender what was good.

"Nothing," the pretty girl answered, "but a lot of people order the fish and chips."

Riffy nodded, "And a beer."

Riffy used a mirror behind the bar to look around, but he didn't recognize anyone. He could see almost the whole room, an advantage because he didn't want to look like he was trying to make friends there—just the opposite. When his food came, he ate it and ordered another beer.

"You from around here?" the bartender asked, looking at him with a slight turn of her head, her voice sultry. Riffy looked up but then averted her eyes, avoiding an answer. He had to play the loner with everyone.

It was after 9:00 before he saw his targets. He walked to the bathroom and then back to the bar glancing at the mirror to see if they had noticed him, but the angle was wrong. Next time he would sit where he could monitor them better —about two seats over. He finished his beer and left.

Back in his hotel room, he reviewed his directives and project plans and sent Chops an email with some refinements and suggestions. Five minutes later, Chops sent him the okay, and Riffy curled up in bed and slept for twelve hours.

The next evening, Riffy arrived at Milk and Fish, just as the ex-cons were climbing out of a taxi. He darted into the bar, but there were no open seats anywhere, least of all at the seat he had chosen to observe his targets. He strode to the booth in the back corner where the cons had sat the night before but stopped short when he saw the "reserved" sign on it. Sweat beginning to stand out on his forehead, he spotted a couple leaving a table nearby and sat down as casually as he could manage, pushing their plates to one side. He positioned himself so they could see his profile and he could see their table in the corner of the mirror. He took a deep breath to focus on the mission.

His targets went straight to their booth, a waitress bringing them their drinks—two mai tais with little umbrellas in them. Riffy snickered. When the waitress cleared off Riffy's table, he ordered whiskey. Glancing in the mirror at the ex-cons' table, they seemed to be looking his way and talking about him, but he looked straight ahead.

One of them scuffled over to his table and stood behind him, saying, "Hey, Hunker!"

Riffy continued to stare straight ahead, his hand wrapped around his glass, but then turned around slowly, allowing his face to contort into a smile and said, "Shit, Bigass, is that you?"

Bigass showed a gap-toothed smile and said, "What the fuck," and turning to the other con yelled, "Hey, it's Hunker."

"Hunker!" the other guy shouted. "Get over here!" Riffy got up slowly and sauntered over to their booth.

"Dickwad!" he said, slapping the man on the shoulder. Riffy felt uncomfortable but also oddly comfortable.

"What the fuck are you doing here, Hunker?" Dickwad asked.

"Just hangin'," Riffy replied.

Riffy ordered a round of drinks, followed by Bigass, then Riffy again, then Dickwad, and by that time they were all reeling—except Riffy, who had forced himself to vomit a couple of times so he wouldn't get too drunk.

After a couple hours of everyone complaining about "the man", Riffy let it slip that he had a big deal going. He had come to New York to celebrate the success of his AI project and was going to make millions. He swore them to secrecy, semi-faking a drunken slur and went on to tell them about Andy Inch and the bank swindle.

Riffy made another trip to the restroom to throw up and Dickwad and Bigass started salivating about bringing Riffy into their organization. This might be just the thing to get them noticed. When Riffy returned, Dickwad said "My boss might be interested in you, Hunker. Maybe take this Andy Inch thing off your hands for a big payday."

"Yeah, whatever," Riffy slurred.

Riffy's foot was in the door. He staggered back to his hotel room, his head spinning from the alcohol he hadn't been able to vomit out of his system. He emailed a semi-intelligible status report to Chops and fell backward into bed, asleep before his head hit the pillow.

He was awakened a few hours later by a banging at the door, but before he could answer, the hotel manager opened it and let Dickwad, Bigass, and another man into his room. Riffy lay naked in his bed, the sheets and blankets in a tangle at his feet.

"Fuck," he said.

"Give me your phone," the new man asked politely. Riffy pointed to his phone, which was charging in the corner. "Where's your computer?" Riffy pointed to the other corner. Bigass grabbed them both.

Riffy had already put his electronics in a locker and sent the key to Chops, so everything in his hotel room supported his story.

"I wanna see what you've done," the boss began (as Riffy had surmised that this is who he was). "These two jokers tell me you've got something we might want."

Riffy, untangling the sheets and covering himself looked at the boss and said calmly and confidently "Let's go to Iowa and I'll show you."

They were on the next plane out.

***

Edie had been preparing ever since Riffy left and had set up an alternate site to keep tabs on Andy Inch. The primary site was ready for the show just a half mile down the road.

When Riffy and his "guests" touched down, they headed directly for the original development site.

"Kind of a dump," the boss grumped, looking around, and Bigass and Dickwad nodded dutifully. Riffy confidently motioned them to a room in the back.

When they sat down, Riffy crossed his arms and lowered his voice. "What I've done here isn't for sale," he said. "I want a piece of the action and a position in the organization. And you can't have the source code. It's my only guarantee that I can stay alive." Riffy sounded more confident than he felt. "After you've seen what I've done, you'll want it, and I know what that means."

The boss nodded, showing a glimmer of respect, shrugging, "Let's see it."

Riffy began with the background on his target bank, followed by an overview of the AI loan officer he had developed. The boss watched intently. Riffy showed him Andy Inch's employment interview with Robert Baron and several subsequent communications, but when Riffy showed him the status screen for Andy Inch's approved loans, the boss leaned forward. Andy Inch had authorized $9 million in loans on properties worth only $4 million. He had electronically falsified the paperwork that

86

showed the value at $12 million, so the bank stood to lose $5 million, which was going right into Riffy's pocket. Riffy told the boss that he had another four banks lined up when Andy Inch finished with this job.

The boss held up his hand, indicating he'd seen enough. "Normally," he said, "I'd try to downplay what you've done to get a better deal, but holy shit!" He turned to Riffy.

"You got a passport?" The boss asked and Riffy nodded.

Riffy suggested dinner before they headed back to the airport and assured them that he knew the best pizza place in town. When they arrived at the restaurant, Riffy went to the bathroom and, as planned, Edie was waiting in a stall, door locked, standing on the toilet. When Riffy burst in, she jumped down and they hugged, holding back sobs. They didn't speak. Riffy kissed her and went back to the table.

At the airport, the boss gave Fatass and Dickwad $50,000 and told them he would mention their names to the big boss but then leaned toward them whispering that if anything went wrong, they were dead.

When they landed in New York, Fatass and Dickwad went back to Brooklyn, and Riffy and the boss went on to the international terminal. Riffy stole a look at the boss's passport and wondered if he should call him Max but decided not to get too friendly.

They spent the next few hours in the Italienne Airline executive lounge. Max had too many drinks to count.

"So what's next?" Riffy asked.

"You'll be fine as long as you're valuable to us," Max slurred. Riffy smiled like he was satisfied with the answer, but his stomach churned. "You gotta make me look good," Max added, his eyes closing.

"You got it," Riffy said.

Max slept through the flight to Rome, but Riffy's eyes stayed wide open.

A car picked them up when they landed in Rome, and they began the two-hour ride to Assisi. On the way, Max shaved and combed his hair in a mirror the driver, Mitsy (who looked just like Max) had thrust at him. "Max!" she said, squinting at him in her rear-view mirror. "You need to comb your hair like I showed you."

Max changed his hair a bit, admired himself in the mirror, and asked, "Better, Mitsy?"

Mitsy didn't answer and tore the wrapper off a candy bar, taking a big bite and quickly gulping it down. She opened another and asked Max if he wanted it.

"Yeah, I'm hungry."

Mitsy ate half the candy bar before she passed it back to Max, who ate it slowly. Riffy wondered who the real boss was but soon he had nodded off.

Mitsy drove them down a narrow street with shops and apartments on both sides. She slowed as they came to two wooden gates, speeding up only after the gates opened and then screeched to a halt at the entrance of a large building: it was like entering a palace in the middle of a dirty little city, as they must have knocked down twenty buildings to build the estate. Max struggled to lift the hefty bags out of the trunk and Riffy was barely out of the car when Mitsy sped off without saying a word.

"That's my sister," Max said. Before Riffy could respond, Max added, "Look, I know you're talented and smart and you have something that works now, but more than that you've got what it takes to be valuable in the future. My bosses aren't as far-sighted as I am, and they would as soon slit your throat as eat a bowl of pasta." He looked directly into Riffy's eyes and said resolutely. "I think you could be vital to what we're doing, so I'm going to back you."

Riffy felt a little guilty that his infiltration into this organization would likely cost Max his life, but the greater good should prevail. Riffy nodded and said, "If I get a chance, I'll have your back, too."

"My bosses aren't the masterminds. It's Freddie in the Mexican cartel," Max said.

"Okay," Riffy said, "good to know," but he had to hide his astonishment that Freddie, the greeter that Remi had first met in Cozumel, was that important to the organization. He needed to meet Freddie and get on the inside track.

Mitsy was already walking up the steps and beckoned Max to get moving. After banging through the front door, she hurried them into a dining room and told them "Sit!" There was food everywhere, but Mitsy stared at Riffy, pointing to the bounty saying, "That's not yours!"

When the bosses came in, Riffy picked up a sandwich and took a large bite. Mitsy looked horrified, but Riffy had to show he was confident and unafraid. Max took the bosses aside and talked to them quietly for several minutes, gesturing authoritatively and pointing occasionally to Riffy.

One of the bosses strode over to Riffy and loomed over him. "Stand up," he ordered. Riffy stood up, inching even closer to the boss who stood directly in front of him. "Max vouched for you," the boss continued, "and you check out so far." He looked Riffy up and down critically and added, "I don't understand this computer shit, but I understand people." The boss stared into Riffy's eyes, and Riffy stared back, putting in his mind that he could just as soon punch him in the mouth as look at him. The boss reached for Riffy's arm, but Riffy knocked it away and continued to stare.

"You can kill me or keep me—your choice, but don't touch me," Riffy spat, continuing to stare.

The boss broke the stare and turned to Mitsy, "Set up the meeting with Freddie."

Riffy was in.

Chapter 19 – Everybody Gets a Necklace

Moto and team catalogued the threat level for each AI-development project, however many of the AI programs were developed in a modular fashion, which made their analysis more difficult. Moto had asked Ginger and Ginger to come up with a strategy for deconstructing an AI program into a personality so that Sophie Jean could analyze them the same way she would for candidates and employees. The Andy Inch program that Riffy had developed became the first test.

Human psychoanalysis uses past behavior to determine future behavior. Has this person lied before? Has he been immoral? For the AI programs, though, the Gingers looked at what a program had the capability to do in the future, so the question was "Can it lie?" "Are there rules in its programming that require it to follow a moral path? "Can it change its own moral compass?" Even the simplest AI program can learn and adjust, and without boundaries and constraints, might evolve into something dangerous.

Moto called a team meeting in the safe room. The team, minus Riffy, sat patiently waiting for Moto. When he came in, Moto handed out what looked like necklaces with pendants to each team member and asked everyone to put them on, saying, "Start the presentation." With that, a holographic image of a blank screen appeared in front of each team member. Moto said, "AI Morality," and his words appeared as the title on each person's personal holographic image. "What keeps an AI program from being dangerous?" Moto asked, but nothing changed on the holograms. The room was silent as the team members exchanged confused glances.

Finally, Hurley said, "Access to weapons," and his words appeared on everyone's screen as a blinking white phrase that, after a few seconds, turned red. "What the hell does red mean?" Hurley asked.

A paragraph appeared on the screen that read: "Access to weapons can increase the impact, but does not in and of itself constitute a threat, as the ability to use weapons is required." After a few seconds, the original presentation screen returned.

"Pause presentation," Moto said, turning to the team with a mischievous smile. "I thought you might like to see what Ball and Chain developed for our clients. This should put another two billion dollars in our coffers." Moto went on to explain how the necklace was a combination holograph and microphone tied into the presentation leader's necklace. The pendant projected a clear image by analyzing the wearer's retina, eyeglasses, and contact lenses while using speech recognition to display words spoken into its microphone.

The leader's necklace controlled the presentation but if someone spoke, the words appeared on everyone's hologram, along with an analysis based on parameters preset by the presenter. So when Hurley said "Access to weapons," the parameters determined that access to weapons might not be a threat because the AI program may not have the capability to use them.

Lily said, "White means the statement is being analyzed. Red means it's of no consequence - sorry, Hurley."

"But what if the technology can't determine an outcome?" Moto asked rhetorically. "Then it turns yellow and says 'let's discuss.'" That was the team's job: to come up with new ideas.

Everyone talked at once, throwing out ideas in an attempt to get one to turn yellow. The software analyzed all of the statements, but they all turned up red or green until Moloana said, "Knowing when to stop." Moto perked his ears. The statement turned yellow.

The room exploded into applause and Moloana smiled. She had raised a scenario that Moto, Sophie Jean, and the Gingers hadn't considered. The team went on for another hour until everyone went silent, but no more statements went yellow. Moto hadn't expected a new idea to come out of this meeting, but Sophie Jean had insisted on trying. He was grateful that Moloana had proved him wrong.

The team learned that AI programs had morality, but it was a different kind of morality. Humans are taught to do the right things and punished for doing wrong, while AI programs had to be permitted to do right things and prevented from doing wrong - effectively putting them in jail before they committed a crime. Computer programmers are usually focused on making their programs "do" something, but for safe AI, the programmer must also tell the program "not to do" things.

After the meeting Chops grabbed Moto and asked, "Which team in Ball and Chain developed the necklaces?"

"Branco's team," Moto replied.

Chops nodded. "Great stuff - what's it going to be marketed as - the Collar?"

Moto smiled broadly, and said, "Calling it the Collar will increase sales by 20 percent."

"So is there any limit to the number of people who can brainstorm in a presentation at the same time?" Chops asked innocently.

"No, except when you have more people, the system slows down." Chops looked at Moto. Then Moto said, "We can fix that - if we had multiple presentation processors to handle the workload, we could literally have thousands of people brainstorming in a single session." He hugged Chops, lifting him off the ground.

Chops disengaged stiffly, content that he had passively manipulated Moto to "come up with" the multiple processor idea. Moto called to Chops as he walked away, "Chops, Riffy needs the information we discussed in today's meeting."

"I'll make it happen," Chops replied without turning around.

***

Chops flew to Iowa the next day to meet with Edie. She picked him up at the airport, wondering if she had done something wrong and whether Riffy was okay, but once Chops hopped in the car, he asked, "Do you have a valid passport?"

"Of course, honey," Edie replied. "I'm always ready to fly." Chops told her about the presentation that the core team had seen and that Riffy needed to see and understand it, but Edie didn't fully grasp what Chops wanted until he stopped talking.

"Oh!" Edie screamed, the car swerving suddenly. "I get to see Riffy!"

Chops put a steadying hand on the steering wheel and said smiling, "let's come up with a plan," and over the next few hours, they developed it, but, seeing the whole picture, Edie became even more frightened for Riffy.

Chops read her like a book. "You know, Edie," he said, patting her knee, "Riffy is on an important assignment, one that may be critical to the future of civilization. You are a professional and need to keep your emotions in check. You can do it, but I can't let you see Riffy if you can't keep it together."

Edie took a deep breath. "You're right," she said, her voice calmer. "I have to keep everything in perspective." She straightened in her chair. "Humanity first, me second," she laughed.

Chops knew she meant what she said, even though it was a joke. He sent a communication to Riffy: AE2. Plan A, E - Edie was the contact, and it was in 2 days.

Edie left for Rome that night, excited, terrified, and determined. Andy Inch had earned a week of vacation at the bank, so she had him take a few days off so she didn't have to worry about being out of the country.

***

Moto had learned an important lesson from the brainstorming session: He was fallible, and he was going to have to rely on his team. Sophie Jean convinced him it was a good thing, and Moto remembered that all his analyses on the threat from AI had resulted in failure. Now he had some hope that what they were doing could succeed even if he couldn't see how. Moto was uncomfortable with doubt. AI wouldn't waste a nanosecond on doubt, but could exploit it in a human. "Wow," he thought. "This opens up a whole new line of defense – exploiting AI's inability to doubt itself."

***

Lily called Moto, but he didn't answer, and she was frustrated about not being able to see him. Lucy called Otis, but he didn't answer; Otis was Otis and he had a lot on his plate. Remi felt trapped and needed someone to talk to. Hurley sat alone at home wishing he could have a beer with Riffy. Edie sat in her hotel room hoping to get a call to pick up Riffy. Beethoven wondered if he would get another assignment out of the office as he rifled through billions of pages of information. Indy got a feeling that something was about to happen but he couldn't decipher whether it was good or bad.

Indy's grandmother, Bohdi, started to cry.

Chapter 20 - Freddie

Riffy awoke from a deep sleep, unsure of where he was, and a hand caressed his shoulder. The touch was feminine, soft, nice.

He opened his eyes and turned his head. It was Freddie. He wasn't supposed to know who she was, so in his sexiest voice, he whispered, "Who are you?"

Freddie just smiled, her lips parting ever so slightly. Riffy had seen pictures, but they didn't do her justice. "Time to get up, tiger," Freddie purred, rolling over and heading for the door, "we have a lot to discuss."

After she left Riffy showered and dressed and checked his electric razor, which flashed "AE2." He was excited to find out that he'd be seeing Edie, but was surprised that that was the plan. He pressed a button on the razor and the message disappeared.

Shaved and ready, he opened the door and saw Freddie standing there, smiling.

"Have you been here the whole time? Sorry—I didn't rush."

"They told me you were ready, and I came back," Freddie whispered throatily.

They? He thought. Now Riffy was certain he was being watched, but he didn't remember doing anything suspicious, even with the razor. They walked a few steps down the hallway and entered a room where two other people were waiting.

"Sit over there," Freddie said, pointing at the head of the table and she sat at the other end. "Okay, talk," Freddie said in a low, breathy voice. For forty-five minutes Riffy described without interruption the original idea, the coding methodology, and how Andy Inch had been implemented. When he paused to reach for a bottle of water, the others peppered him with technical questions about the video interface, how the Andy Inch brain worked, and how Riffy controlled his AI program. Riffy replied with technical answers but held back on some key technology he had developed, giving them just enough to look transparent without giving everything away. The hours rolled on as he showed them a video of the Andy Inch employment interview, confidently answering their technical questions.

Freddie hadn't said a word, but once her team ran out of questions, they looked at her and shrugged, so she tossed her head towards the door and they walked out of the room.

Riffy turned to Freddie. "Do you have any questions?" he asked.

"Yeah, tell me what you're hiding."

Remembering the character he was playing, Riffy stood up and unzipped his pants. Freddie laughed. "Hmm," she said, "come on, big boy." Freddie grabbed Riffy's belt and pulled him out the door.

She told the guard to bring dinner, leading Riffy along the corridor and then into her room. "Sit over there," Freddie said, pointing to a long, white velour sofa.

Riffy plopped down and tried to look comfortable and confident, but he was worried about what might come next. Freddie slipped into an adjoining room and slinked back in wearing a gray silk robe that flowed to the floor. She sat close to

Riffy, crossing her legs and leaning close. Then she said the last thing Riffy expected.

"I'm dying," she said, "and you're going to replace me."

Riffy just stared at her, unable to speak. Looking into her eyes, she was telling the truth, and he felt compassion, attraction, shock, and fear all at the same time.

Attraction took over.

***

Riffy woke up early the next morning and crept out past the guards to his room. His mind struggled with guilt for cheating on Edie, compassion for Freddie, satisfaction for getting into the organization, and fear that he might not make it out alive. He wished he could be with Edie.

Plan A with Edie was no longer possible, so he went to the bathroom and sent a "C" to Chops on his razor. A few seconds later he got a CE1 back confirming plan C, Edie, in 1 day.

After Riffy showered, shaved, and had breakfast, he waited on the couch for Freddie to show up. When she arrived, instead of sitting with him, she beckoned him down the hall to a large study. A tall, blonde man with a sweet face entered behind them.

"Hello, Riffy," the man said, holding out his hand, "my name is Henry." Riffy shook his hand, glancing over his shoulder at Freddie, who had retreated a few steps. "Welcome to the Gulag," Henry winked.

Henry darted to a table at the side of the room and poured a drink. "I'll be open with you, Riffy," he said, "I don't want to be here, but this organization has threatened my family. And don't think they're not listening to our conversation either." A sad look came over Henry's face as he looked down briefly and then turning back to Riffy, explained, "as long as I do well here, my family is safe." Riffy was quiet as Henry continued. "As you know by now, Freddie has stage-four kidney cancer and may not be with us much longer. She's the technical brains and I manage the development."

Riffy didn't know that Freddie had kidney cancer, but it didn't matter what was killing her. Henry must have seen Riffy's eyes widen, as he said, "Sorry, Freddie. I thought you'd told him the details." Freddie shook her head like it wasn't important.

Henry plopped heavily into a small metal chair causing it to screech and continued. "Since you're a permanent part of this organization now, we're going to bring you up to speed on the design." *Permanent part of the organization,* Riffy felt a gnawing pit in his stomach but stayed in character, keeping his composure.

Henry was clear and concise as he explained what RICKY was trying to accomplish. Riffy asked a few questions about how far they had gotten, but Henry ignored them, focusing on the design and purpose of RICKY's AI development. Riffy was both impressed and unimpressed with what they were doing. There were some things that seemed just stupid and others were brilliant – *probably the influence of the mob bosses,* he mused.

Freddie sat quietly, distracted, next to Riffy. She had no relatives, but she had an adopted family she loved. RICKY had promised that her first and best friend Calliope would be taken care of for the rest of her life.

Freddie was orphaned when she was just six years old and wandered the streets of Houston, scraping for food. One Christmas, she crept up the steps of a pretty little house in Lindale Park and rang the doorbell. When the door opened, she walked right in, but instead of calling the police or child services the couple that lived there, prodded by their grandmother Calliope, accepted her into their loving arms. Calliope took Freddie under her wing and her grandchildren followed suit, falling in love with Freddie and raising her with their other children. Freddie owed a debt to Calliope and she was determined on repaying it, literally on her deathbed.

At the end of the long day, Riffy and Freddie went back to her room, had dinner, and spent the night together. The following morning, Riffy suggested they go to the Improvisario Restaurant for dinner. Plan C was for Riffy to meet Edie in the men's restroom there, but as the day went on, Riffy saw that it wasn't going to happen, but since Edie would be there for five nights, waiting for him to show, there would be other opportunities. Riffy let Chops know that it was going to take a little longer.

Freddie was tired again, so at the end of the day, Riffy went back to his room and Freddie went back to hers. Riffy loved Freddie in an incomprehensible way, but his love for Edie was clear – he just hoped Edie would understand.

The next day was long and arduous - they continued to work on the transition. Freddie was even more tired and begged off going out to dinner. Shaky, Riffy was worried about her, so he spent the night with her again. When he awoke, the radio was playing Rachmaninov's "Prelude in C Sharp Minor," and Freddie was entwined in Riffy's arms and legs.

She sighed and looked up at Riffy, "please take care of Calliope," then she closed her eyes and breathed one last breath. The piano concerto continued as Riffy held her close, trying to keep her cooling body warm. A tear slid from the corner of his eye and into her hair.

There was a knock on the door. Riffy arranged Freddie's gray silk robe over her before he answered it. Henry stepped into the room and Riffy leaned his head on Henry's shoulder and shuddered. Henry patted him on the back and told him to go back to his room. Riffy walked slowly and deliberately to his room as a raft of people hurried past him.

He had only one more night to meet with Edie, so he turned his fractured mind to finding a way out of the compound. He sent "B" on his razor, and Chops responded immediately with BEO, a meeting tonight or tomorrow. Riffy ordered coffee and breakfast but couldn't stomach any of it. He wanted to contact Calliope but he didn't know how. He wanted to talk to Edie but their meeting was many hours away. He was tired and scared.

***

Riffy woke with a start, hearing a knock on his door - he hadn't realized he'd fallen asleep, but his bedside clock told him it was already early afternoon.

Henry took him to a large team meeting of the AI development staff, where he announced that Freddie had passed away and that Riffy would be taking her place. There was little reaction.

Remi was in the room. Riffy addressed the team and sent a coded message for her, making brief eye contact as he spoke.

Riffy got status from each team leader and went to see Henry. "You want to go out for a drink tonight? Riffy asked, rubbing his eyes.

Henry smiled, "I think we could both use a little escape."

"You ready to go now?"

"Give me an hour. I'll meet you at…?"

"Improvisario Restaurant," Riffy replied as casually as he could manage.

Riffy had the guards call a cab and grinned when he saw the curly red hair of the cab driver. He bounded down to the cab two steps at a time, got in, and slammed the door as Edie took off slowly, trying to milk as much time as possible from the cab ride.

The drive was only twenty-three minutes, so they had to work fast. Edie handed Riffy a Collar and he reviewed the presentation. That took ten minutes. For the rest of the ride, Riffy briefed Edie about the people he was working with and about how RICKY was developing virtual programmers and virtual analysts. He demanded that Edie take care of Calliope – Freddie's dying wish.

Riffy's hand lingered on Edie's as he tipped her for the cab ride. Already lonely, he walked in a daze towards the restaurant.

Chapter 21 – The Collar

"For Thine is the Kingdom and the Power and the Glory Forever. Amen." Moto lay in bed looking at a photograph of Eli while listening to a recording of Andrea Bocelli singing his iconic version of the Lord's Prayer and thought about his grandfather, hoping he had this song in his heart when he slipped away.

Moto's fractured mind prevented him from concentrating. It happened sometimes. He usually kept everything compartmentalized, but Moloana's new idea in their brainstorming session had caused him to re-evaluate everything he had filed away so carefully over the years.

Everything is so fricking interconnected, Moto thought, but as he had done countless times before, he applied Moloana's new idea to each area–the doomsday sites that Moloana and Lily had set up, the analysis routines that the Gingers were using to identify AI development, Sophie Jean's psychological profiling, Hurley's background checking, Lucy's anti-AI programming for businesses, the electrical grid quantum programs and their inter-grid communications that Otis and Indy— now Weetzie— were working on, his own ideas for his Secret Algent program that he hadn't full developed yet, and his dream communications with Lily. One by one, Moto re-analyzed each environment as if he were stepping through computer code. It took days to document the changes and send his analysis to the team *and* to the lockbox in his brain.

Chops was responsible for assuring Moto was kept up to date. Thanks to Sophie Jean, Chops understood how Moto compartmentalized his thoughts so when communicating, he targeted just one of Moto's mental compartments—Chops called them MMCs— at a time. It took Chops months to master, but it helped keep Moto in balance and communications efficient.

Chops used an app when updating Moto that conferenced in up to twenty people at a time, although they seldom needed more than one or two, as the calls were limited to answering one question at a time. This process helped Chops remain in control of the meetings with Moto.

Chops had been home for only six days in the last eleven months. His responsibilities required him to travel the world, mostly for the interviews he conducted, but he had directed the team to expand the Collar technology so they could connect through the electrical grid, allowing Chops to do his presentations from anywhere in the world without a phone. Otis had taken it a step further by inserting a lip-reading module and facial recognition into the Collar. All Chops needed to do was mouth the words and his voice transmitted to whomever he was speaking to. If Chops started a sentence with "Moto," only Moto would hear what he had to say, and if he started the sentence with "Sophie Jean and Moto," only those two would hear him.

The team developed an app for the Collars called the Leash, which controlled who could have private conversations during a presentation. Chops had a variety of ways to control that, including one he called the Choke Chain that would stop any unwanted chit chat immediately. He didn't want anyone but Moto to interrupt him, so he and Moto had equal priority, and their voices could be heard simultaneously. Others on the call were heard only if no one else was speaking.

For Chops' first meeting with Moto using the enhanced Collar, he had sent them to everyone who might be pulled in to answer a technical question. He had tested the enhanced Collar with each team lead and was satisfied with the results.

The Collar that Ball and Chain marketed used a straightforward wireless interface to the internet, but that wasn't good enough for Moto's team because their conversations had to be undetectable. A small device plugged into what looked like a pinhole on the Collar connected to the electrical grid, allowing for complete privacy. While he showed no emotion about it, he was bubbling with excitement about what the team had done.

Chops called Moto to start the status meeting and Hurley walked into Moto's office handing him the enhanced Collar, and when he put it on, Moto saw Chops' smiling face on the hologram.

"Happy Birthday," the holographic Chops said with a big smile.

"It's not my birth—" Moto began. "Oh, wow, it is my birthday!"

"And I have a birthday present for you from all of us," Chops said. This is a secret new technology we incorporated into the Collar strictly for our team's use." Chops went on to explain.

As Moto listened, pride in his team for inventing the new technology was dampened with concern that he was a year older and had a year less to save humanity. When Chops finished his explanation, Moto was overwhelmed but managed to choke out, "thank you, Chops."

Chops stammered an unintelligible response, clearing his throat, and said a little huskily, "Let's get down to business." He put the presentation up on the Collar and went through the team's status reports and plans for remediation of Moloana's idea.

As the status meeting continued, Chops brought several team members into the conversation to answer questions, and the Collars worked flawlessly. A few of the team members just mouthed the answers to their questions because they didn't want to disrupt the people around them, but the Collar technology converted their lip movements into their own voices. Chops explained that the videos Moto was seeing were only simulations of each person because actual video transmissions might be too burdensome for the network.

When the meeting ended, Moto thanked Chops again and left to thank Otis and Weetzie personally. When he strode into Otis' office, the whole team was there with a birthday cake, singing an English baroque version of the late 1800s "Happy Birthday" song that has become so familiar. Moto just shook his head and smiled broadly at his team. An image appeared in his head of Indy smiling.

Chapter 22 - Beethoven

Beethoven had been concentrating on data mining for the past few months while Otis worked on new technology. Beethoven adjusted the SNIFFER app to discover devices that were daisy-chained rather than plugged directly into the outlet, resulting in an additional 4.5 billion devices worldwide that they could monitor, most of which were computer-related.

He noticed patterns - a grid might go offline then return, devices disappear and then come back, and some grids, usually in developing countries, went offline frequently. Beethoven established parameters for these anomalies. He noticed that some devices would appear on different power grids or locations each day, usually explained by people taking their laptops home at night and back to the office the next day. Beethoven focused on laptop activity - he didn't know what he would find by monitoring it, but he had a gut feel that one day it would be important. Otis told Beethoven that he didn't see much value in it, but left Beethoven to do his own thing.

Beethoven programmed the worldwide monitoring board to highlight abnormal conditions, painstakingly reviewing each one and adjusting his routines accordingly. For eleven weeks, he honed his analysis until no anomalies showed up. He was a little disappointed that all of his work had not found any real problems - yet.

Otis walked up behind Beethoven, who was looking at the monitoring board. "I don't see any alerts," Otis said sarcastically.

Beethoven began, "no, I've iteratively adjusted the analysis to allow for reasonable and rational —"

Otis interrupted, "Looks like all this effort was a waste of time. You haven't found any problems," he said tongue in cheek.

"I think it's important," Beethoven responded defensively.

Chuckling, Otis said earnestly, "Beethoven, I was kidding. With all the work you've done, we'll be able to see when there really is a problem," but then after a few seconds, Otis said, "but so far, you know, you got nothin'," and he howled. Beethoven smiled, his shoulders relaxed, chin up, glancing over his shoulder at Otis.

Beethoven monitored the devices Riffy and Remi were using closely, along with the rest of the team, including their Collars. They stopped using their phones, as the Collars were easier and more secure. Personalized with voice and facial recognition, the Collars connected to the electrical grid network, secured from hacking and eavesdropping. Otis added an interface to recharge the batteries automatically, and rather than using GPS to track location, Beethoven could monitor movements through the electrical grid network, providing yet another level of security.

Otis was most proud of the sound technology, which he called Floppy Earphones, creating sound waves that were focused, narrow, and directed to the wearer's ears, incapable of being overheard. It sounded like a clear whisper.

The Collar that Ball and Chain marketed to the general public cost a couple hundred dollars apiece to manufacture, but each Collar that the team wore each cost $1.5 million. Moto and Chops felt it was justified, so Moto had forty Collars made and distributed to key members of the team. He would eventually need more, but the technology would improve as time went by, so he kept another sixty waiting in the assembly line.

Beethoven was proud to have received one of the Collars; he wore it every day and set up alarms to come to his Collar from his data-mining activities. Otis kidded Beethoven that he was too anal about his alerts, but Beethoven just sloughed it off and remained diligent.

At 2AM one morning, Beethoven began getting alerts from Remi's and Riffy's computers and the devices on the electrical grids in Assisi, Italy. He alerted the team, whose Collars came online immediately.

Moto was first to speak: "What's going on, Beethoven?

Beethoven tried to sound calm as he told the team that something was happening in Italy, but he wasn't sure what. Riffy and Remi's devices had moved, and many of the devices were no longer online.

"What do you think it is?" Moto asked.

Beethoven reeled in his anxiety but turned to the data associated with the alerts. A few of the devices on the original grid had moved to a new grid that was about 175 kilometers away but most had gone offline.

Tick – tick - tick – Beethoven felt the pressure, as though he could hear Moto drumming his fingers. "Moto," he said at last, "I believe RICKY is going into systems testing. The large development site is shutting down and a smaller site expanding." Beethoven told the team, waiting for a response.

After a few seconds, the team members who had been getting the spotty statuses from Riffy confirmed that Beethoven's analysis was consistent with their perceptions of what was going on in Assisi.

"I should have seen it earlier," Chops said, "good job, Beethoven."

"Yeah, great job, Beethoven," Moto said. "Team, you know what to do. Chops, keep me up to date."

Chops nodded, and the Collar sent a visual of Chops saying "Will do" to the team. Chops would never say such a thing, but the Collar was programmed that way. Moto sent a note to Otis about it, joking about his concern for the Collar's programming.

Otis "Collared" Beethoven after the meeting to apologize for not being involved and supportive of what he was doing with data-mining. Beethoven responded "no worries", understanding that Otis was drawn to new technology and proud that he was getting some much deserved respect for his efforts. As point man for data-mining, he had made a name for himself, and was now essential to the overall effort.

"We need a backup for me, Otis," Beethoven said with more self-assurance than he had ever heard from himself.

Otis nodded. They did.

Chapter 23 – Edie Keeps Her Promise

Edie felt like she was in jail. She wanted to be with Riffy. She wanted to find Calliope to ensure Freddie's last request was met. What she didn't want to do was babysit Andy Inch as he robbed the BOD. She collared Chops to vent her frustrations and even though Chops was not a psychologist - Edie didn't really need psychoanalysis – he understood she needed a plan for how to get back with Riffy.

"We can get this done, Edie," Chops said. Much of what you want is already planned. I need two days, and I'll tell you exactly how it happens. You'll need to be ready."

Edie blurted out. "Ready for what, Chops?"

"Anything!" Chops replied.

Edie calmed down. "You're the tops, Chops," she said with a grin.

Chops arranged to go to Houston and to have Hurley meet him there. The mob had been paying for Calliope's care in an exclusive River Oaks home, and Chops and Hurley figured that these payments would end once Freddie was gone, so their first stop was at a mansion on River Oaks Boulevard. When they pulled up, an ancient woman was being led to a car, her steps slow and unsteady.

"Now, now, Calliope, everything will be fine," Chops heard the nurse say and he jumped out of the car and ran to them, waving his arms over his head.

"There's been a mistake! Take her back," Chops shouted, arriving breathless at the nurse's side. He slipped her two hundred dollars and followed the two of them through the oversized wooden front doors, quickly identifying the person in charge and taking him aside. "Calliope's payments will continue," Chops told him, "but they'll be coming from a different source."

"It's $90,000 a month," the man said, looking more than skeptical. Chops opened his briefcase and gave him the next month's payment in cash. The man nodded as if this happened every day. "Yes, sir," he said crisply. "Would you like to visit with Calliope?"

Chops nodded, and the man led Chops and Hurley, who had joined them, to Calliope's room. When they got there, Chops introduced himself while Hurley did a quick sweep - as expected, there was no surveillance.

"I'm here on Freddie's behalf," Chops told Calliope, his voice quiet. "She asked a friend of mine to look in on you and make sure you have everything you need."

"She's gone, isn't she?" Calliope asked. Her voice was considerably stronger and steadier than Chops expected. "She missed our scheduled talk last week and they tried to move me out of here." Calliope reached for a chair and sat down with a sigh. "She didn't know how long she had. It doesn't seem fair; I'm ninety-nine years old, and she was only twenty-eight." Calliope gathered herself and added, "Now, tell me why you're really here."

"There's no other reason. Her last wish was to provide for you and give you anything you want," Chops replied.

"I have an older sister," Calliope said after a moment, pulling a lacy handkerchief from her pocket and dabbing her eyes. "Her name is Tux, and I would love to see her–be with her–again. She moved to Fort Worth a few years back." Chops nodded to Hurley, who got up and left for the airport while Chops stayed with Calliope, attempting to comfort her.

***

Hurley found Tux's address at a small assisted-living facility in northwest Fort Worth, while taxiing to the airport. He asked the facility's manager to find out what Tux's needs would be if she were to travel to Houston. Tux was mobile, so Hurley booked a flight for later that afternoon back to Houston for the two of them.

Tux was 109 years old and had been blind for thirty-two years after a radiation treatment she'd received for a hyper-thyroid condition. When Hurley arrived, he tried to take her arm, but she refused, walking to his car on her own, and then when they arrived at the airport, she somehow found her way through security. She would ask Hurley impatiently whether she should turn left or right but otherwise she remained staunchly independent.

The flight was short and there wasn't much conversation. But when they got to Houston and hailed a cab, Tux began to fuss with her purse and scarf.

"Are you okay?" Hurley asked.

"Of course I am!" Tux snapped, and then a little more gently, "I just hope she is." Tux dabbed at her eyes with a starched, ironed handkerchief. This was not Hurley's bailiwick so he decided not to try to comfort her because he thought she might punch him.

Chops tracked Hurley and Tux's progress, and when they pulled up to the front of the house in River Oaks, Chops and Calliope were waiting outside. Tux stepped out of the car unassisted and walked toward them, first bumping into Chops, then Calliope. Tux and Calliope embraced wordlessly, patting each other's backs with affection.

After a few seconds, Calliope said, "We're having whitefish, girl. Hope you don't mind."

"I'm starving," Tux said with a toothy smile. "Let's get this show on the road."

Calliope took Tux's arm as they walked in. "Hey, girl," Calliope whispered.

Chops secured permission from the owner of the River Oaks mansion for Tux to stay with Calliope, and Hurley was to go back to Fort Worth to have Tux's belongings sent there. At 1:00 am, Chops and Hurley finally checked into a hotel and gratefully said good night.

Tux talked non-stop, and Calliope loved it as they curled up on the couch until their evening meal arrived—plates filled with blackened redfish, chicken croquettes, and liver pâté. After their meal, they walked back toward the couch, but Calliope tripped, and they both fell, laughing, in a heap onto the soft carpet. Calliope went silent and stopped moving. Tux could

115

sense the end was near and she caressed Calliope, telling her everything would be all right, talking softly and rocking her until she stopped breathing.

Tux lay her head on Calliope's shoulder, comforted that she had been with her one last time and saw an image of Freddie, smiling, welcoming and next to Freddie was Tux's mother, dressed in white, beckoning. Tux closed her eyes as she and Calliope floated away, arm in arm.

***

Hurley had installed surveillance cameras in Calliope's room, and when he reviewed the recordings the next morning, he showed them to Chops. "Please edit these to show how happy Calliope and Tux were before they passed," Chops said. The resulting video was tasteful and sentimental and showed the love that Calliope and Tux had for each other and their happiness at their final reunion. Edie and Riffy would be proud of what they had done in Freddie's memory.

Freddie had been paid well by the mob; her account had $246 million in it. Chops quickly moved the money nine times, finally leaving it in an untraceable bank account in Belgium. Freddie's money would pay for the plan to extract Riffy and Remi from RICKY.

Chops and Hurley finalized the extraction plan and Hurley left for Iowa to meet with Edie. Chops sent "EX9" to Riffy's razor and Riffy responded with "6," indicating that it would take six weeks to complete his work before he and Remi would be ready.

Remi and Riffy used a code to communicate; Riffy would walk by Remi, and if he said the word "you," the next word's first two letters would be the first part of the code. Then the first number he said would be the rest of the code. As Riffy passed by Remi's desk, he said loudly, "You exactly what I need. I'm gonna send you a half dozen roses." EX6.

Chapter 24 – Too Timpani

Moloana and Ginger became best friends on the Maui trip and started playing the timpani together, at first just for fun, but then they tried a one-night stand at a small bar and the audience loved it. (Of course, it would be hard not to like two beautiful women pounding timpani to a Brazilian beat.) They began getting requests to do more engagements but had to turn many down because they didn't have time - after all, they were on a mission to save the world.

Lily told Moto she wanted to go to one of Moloana and Ginger's timpani gigs and surprisingly, he was agreeable, so Lily took immediate advantage, stopping by his office at the end of the day and telling him they were going that night.

"I'm too busy," Moto said, predictably.

Lily took him by the arm and pulled him out of his chair. ""Timpani tonight!" she insisted.

Moto opened his mouth to argue but the determination in Lily's eyes made him smile. "I'm shutting down now," he assured her.

As excitable as Lily was, she was usually extremely patient with Moto.

They arrived at the venue and were surprised by the long line and that several people had t-shirts with pictures of Moloana and Ginger on them. The bouncer at the door walked down the line to them, escorting them inside, and seated them at a table center stage, right next to the dance floor. A waitress came by and said their order would be out in a couple of minutes, and then turned to leave. Lily and Moto started to say they hadn't ordered yet, but they were interrupted by applause as Ginger walked from offstage towards their table.

"Thanks for coming, guys" Ginger said in a sweet, sexy voice, "you're our guests tonight." And she walked away, smiling as she slinked through the curtains.

Moto and Lily looked at each other with wide eyes and big smiles and their server brought drinks and a pupu platter. (Ever since the Maui trip, they referred to any appetizer as a "pupu.") The lights dimmed and a laser displayed "Too Timpani". The crowd cheered as Ginger and Moloana came out dressed in silver and red sequined outfits, parading slowly into position behind the two timpani. Lily and Moto stared at the spectacle.

The performance began with Beethoven's "Moonlight Sonata," which slowly turned into a rock version of the beautiful, quiet piece. By the end of the sonata, the dance floor was packed with enthusiastic dancers, including Lily and Moto. Moloana and Ginger upped the tempo, featuring songs from Brazil, Africa, and Ireland and for the next ninety minutes Moto and Lily didn't sit down but swung each other around the dance floor, oblivious to the world. Moloana caught Ginger's eye and the two musicians laughed, playing faster and more furiously.

Two Timpani performed only one set, although it was a long one. No one performed before or after them—it was just ninety minutes of non-stop frenzy. When they ended their performance, they jumped to the dance floor, hugging everyone. Many tried to give tips, but they told them to give the money to the servers. When they reached Moto and Lily, Moto smiled broadly, speechless. Lily was able to mumble, "Amazing...," but it was inadequate to express her true feelings.

Moloana and Ginger jumped back up on stage and bowed to the audience. Then they turned, nodding to each other, and went back to the timpani, playing an African drum encore that brought the crowd back to a frenzy. Moto yelled to Lily over the pounding music, "I want to be assigned to the same doomsday bunker that Moloana and Ginger are in. Make sure we have timpani in the bunkers."

"Consider it done," Lily yelled back.

Chapter 25 – Moto's Liver

AI development was growing exponentially throughout the world, and the Gingers needed more manpower to keep up. Before the Collar technology, they would have asked for a meeting in the safe room, but now they could schedule a secure Collar call. Moto and Chops were not surprised when they received the Gingers' request.

Chops helped the Gingers prepare the presentation for the meeting. He was a little uncertain when they decided to add background music of Ravel's "Bolero" to demonstrate how far they had come and how close they were to the climax, but he soon realized this dramatic music would help them communicate the urgency of the situation. Ginger and Moloana recorded a timpani overlay for an even more dramatic effect, and the presentation was timed with the music. The Gingers felt sheepish about how much time they spent on the presentation, but had come to a critical point and needed to keep everyone's attention—and just maybe show off their technical and artistic prowess.

Chops sent an alert for a Collar meeting in eighteen hours, and Lily made a note to be with Moto so he would be on time. It was unnecessary, though, because Otis had added a feature to the Collar that not only forced open a communication to the participants but also tracked whether or not they were wearing their Collars. If they weren't, they were instructed via voice message, email, and text to put their damned Collars on.

Just after Moto received the meeting notice, he got a call from his doctor. The routine blood tests that he did each year showed that he had slightly elevated ALT liver enzymes, meaning liver cells were dying or being compromised.

"But I feel great, doc," Moto protested.

"Good," the doctor began, "but this liver enzyme should be checked out. The first step is to recheck it in a couple of weeks and, if it is still high, do an ultrasound."

"What's the worst case, doc?"

"There are many possibilities, but an ultrasound is the best tool we have to narrow it down."

"Okay, doc, but know this: it is important to humanity that I live a long life."

The doctor paused, "Don't worry: there are a lot of things that this could be that are very minor. We'll send someone out to take your blood in a couple of weeks."

"Okay, doc. Thanks." Moto quickly determined this was out of his control, there was insufficient data, and he refocused on his work.

***

Employees at Ball and Chain were required to give the company's doctor access to their medical records and Sophie Jean received updates from each employee's physicals and doctor visits. As she read through Moto's test results, she began perspiring (only men sweat), thinking about the very serious things that could have caused these results.

Unable to sleep, she analyzed every possible disease that could cause this test result, identifying next steps, tests, therapies, and probable outcomes. She contacted Hurley to hack the national registry of the liver transplant waiting list. Hurley sent Sophie Jean a list of everyone in his database whose blood type was compatible with Moto's, and provided them with organ donor forms. Three people in the Ball and Chain organization were a match, and one was Riffy. "Half a liver is enough," Sophie Jean mumbled aloud before she finally fell asleep on the couch, covered with printouts of all the possible iterations of what could happen.

Hurley was still awake when an image of Indy shaking his head "no" popped into his mind, but once he Collared Lily and Moloana and asked them to set up a secure medical lab in each doomsday site, the image changed to Indy smiling. Satisfied that he had done all he could, he opened a beer and settled down in front of the TV to watch the last leg of the Tour de France.

Chapter 26 - Bolero

Ginger prepared an algorithm plotting the growth of AI development over time on a large spinning world holograph. When she projected future growth, AI saturation happened within eighteen months. The goal of the meeting was to find ways to slow it down, so she categorized all the AI programs into AI personalities, to simplify the communication.

Ginger and Ginger had gotten off to a slow start tracking RICKY because its development was off-grid, but since Riffy and Remi were in place and RICKY's network was finally connected, Ginger got access to the program code and was able to "personalize" profiles of the AI "beings" RICKY was developing.

Just before the meeting, Chops touched base. Ginger was looking forward to the meeting and being the center of attention, but Ginger touched Chops' wrist and said, "I'm nervous! I put all the data together for the slides and I'll have to answer all the questions."

Chops pulled away and gave her a thumbs-up. "You've got this, Ginger. You really know your stuff." Ginger forced a weak smile and her sister cheered her on.

Ravel's "Bolero" is seventeen minutes long, and the Gingers timed their presentation exactly. For effect, the presentation would show the world saturated with AI programming as Bolero's crescendo reached its climax.

Chops opened the meeting and quickly turned it over to the Gingers. A holograph of the western hemisphere appeared showing a few yellow dots and many green dots on the map, representing AI development. The three dimensional map started to spin slowly like a globe. Overlaying the holograph was the date the team first met in the safe room.

"Bolero" played in the background as Ginger summarized the safeguards they had inserted into AI programs around the world. She handed off the presentation to Ginger, who explained how RICKY was developing AI beings that replaced human programmers and analysts and would do the nefarious bidding of RICKY's criminal backers. In addition to the color RED, which was self-explanatory, she introduced a new color, purple, to the world map, explaining that purple represented AI programmer beings. The music stopped and the map showed the current month and year. Ginger said, "Please watch."

The music resumed and the color-coded holograph showed the status of AI development, but as the Bolero 's volume and tempo increased, the world map became increasingly yellow, then red, then purple, and as Ravel's masterpiece reached its brilliant climax, was solid purple.

It was 18 months in the future.

The room went silent until Chops said quietly, "This is our projection. Our job is to stop that."

"Ginger and Ginger," Moto said, standing and using the correct inflection for each of their names, "thank you, this is the scenario I feared when we began. We can hope that these AI developments, like most, will take longer than planned, but we can't count on it. We have to fight this. Chops, please continue."

The brainstorming session began with everyone talking at once and in thirty minutes they ran out of ideas, but there were two coveted yellow ideas, requiring further investigation. Moto smiled at Moloana, seeing that both ideas were hers.

Chops announced that they would reconvene the next day and that Moto and Ginger would determine what combination off actions would result in the greatest success and then he prompted Ginger to continue.

Ginger displayed a map of Omaha, Nebraska and zoomed to different locations in town where police, judges, teachers, shopkeepers, bank tellers, technicians, politicians, and others influenced the community's everyday life.

"These people," Ginger said, "are targeted to be replaced by AI." The team was quiet. "But not all of these efforts are nefarious," Ginger added. "Most AI development projects have good intentions. An AI judge is being taught to look through case histories to determine precedent. An AI policeman searches backgrounds of persons of interest to determine how to approach and question them."

Chops spoke up, "But our task is to identify the most dangerous AI personalities, regardless of intent."

Ginger nodded, continuing, "We need you to help determine each identity's risk to humanity."

There were fifty-three entities to review and as expected, the analysis that Ginger, Ginger, Chops, and Moto had done was exhaustive, and there were no new ideas for the first forty-seven, but the forty-eighth entity was an AI librarian at Iowa State University. When Moloana said, "Read how-to books," her suggestion flashed yellow. An AI librarian that could read and understand "how to do or make anything" could be of assistance to customers, but without moral guidance, a risk to humanity.

Chops reminded the team that they would take a fresh look the next day after they got some sleep.

Chapter 27 – Chops Moves On

"You seem distracted, Chops."

Ever since Chops saw the projection of eighteen months until AI saturation, he longed not to be alone. He couldn't get Daisy out of his head. When he travelled, which was all the time, he would lay over in New York so he could drop by and see her, and she was always willing to meet. They had had several meals together and great conversations, but until now Chops hadn't felt he had time for a relationship, but now he felt an urgency to act.

Chops sat quietly, staring straight ahead in Moto's office, and then glanced at him.

"Go on and see her, Chops," Moto said, laughing. "I'll get Branco to fill in for you."

Chops gasped. *How could Moto know about Daisy? Oh, okay, right, the movements of every teammate are tracked.* Chops hadn't opened up his feelings to anyone, least of all to Sophie Jean, but apparently she had figured it out–probably even before he had.

Branco had been trained to be Chops' backup. He didn't have the insight into people and charisma that Chops had, but he knew how Moto's mind worked and he had the same level of project management skills as Chops did.

"Don't worry, Chops," Moto continued. "If we need you, we'll Collar you."

Chops stared at Moto, and the irregular, toothy smile that he seldom showed flashed. "Adios, amigo," Chops said as he leapt out the door, "and thanks."

Moto called Branco in. They hadn't had much contact before, but Branco was up to date. When he walked in and Moto told him he'd be filling in for Chops for a while, Branco burst into non-stop chatter. Moto was glad that he had the confidence to speak up and be a collaborator, but after a minute or two he held up his hand to stop him.

Branco grinned and said, "Sorry—I talk a lot when I'm nervous."

"No need to be, Branco. Chops has full confidence in you which means I do as well." Branco took a deep breath and sat down as Moto continued where he and Chops left off.

Branco was different. He had many of the same important qualities that Chops had—intelligence, project management, and communication skills—but he didn't have the insight into people to sway them to his point of view. However, he wasn't afraid to confront Moto if he felt Moto was going down the wrong path.

As they continued their analysis of the brainstorming session, Branco sat back in his chair, his arms folded, "I don't see that there's anything we can do that will stop AI dominance in the long run," he said.

Moto sighed as though the weight of the world lifted from his shoulders, "You're seeing the battle the same way I've seen it for the past five years."

"This is how I see it, Moto. Plan A was to control AI development. Plan B is to either sabotage or stop it. Plan C is survival in an AI-dominated world. We just need a plan D."

Moto leaned forward and spoke quietly, almost in a whisper, "You have great insight, Branco. Chops never really came to terms with this inevitability. You've come to the same conclusion I did. Given that though, let's work on plans A and B to give us the most time possible before we have to go to plan C. Hopefully, we'll find a way to un-inevitable the inevitability."

Branco's shoulders relaxed and he leaned back with a look that could only be described as confidently scared to death. "Got it, Moto," he said, "let's get to it."

They analyzed next steps, talking freely about the concern that AI dominance was inevitable. Branco was the perfect sounding board for Moto and although Chops had been invaluable, Moto now wondered whether he might not be the best number-two man for him from this point on. His team was entering a new phase of the battle, and a total reorganization of the team had to be considered. After Branco left, the information they discussed carefully compartmentalized in his mind, Moto contacted Sophie Jean, Chops, and Hurley and asked them to re-evaluate the team's roles in light of the challenges ahead.

***

Chops had just finished a glass of wine with Daisy when he got Moto's request to reorganize the team. He hadn't considered it, but knew it was the right thing to do, and felt he should have thought of it. This new eighteen-month timeline was a shock to him, and preparing for Gingers' meeting was all-consuming, so he gave himself a break. He and Daisy talked about it for hours, running through all the possibilities, but in the end, they agreed that, as an elder statesman, Chops was still needed on the team.

Daisy carefully placed a blanket over Chops who had fallen asleep on her sofa. She looked longingly at him wondering, hoping that he felt the same warmth towards her.

When Chops woke up, Daisy was in the shower, singing a Christmas carol. Chops pulled back the shower curtain and sang, gruffly, "Fa la la la la la la la laaa." Giggling, Daisy pulled him in.

After a long shower, Chops lay on the bed thinking how happy he'd been working with Moto. But with Daisy he felt complete and he liked that feeling more. *Aaaaah, yessssss*, he thought.

Chapter 28 – The New Organization

Edie took charge of the RICKY extraction plan, determined that Riffy, Remi, and the world would be saved. The project had unlimited resources; the $245 million from Freddie's bank account was sick, but Edie wasn't extravagant. She didn't want to give a dime to RICKY that she couldn't get back. "Hell, Edie," Hurley said when he looked at the changes Edie made to the plan, "you could've been a grifter ... if you didn't have those pesky morals."

***

Chops, running late, began the meeting as if he weren't lying in bed with Daisy. "So our two objectives," he announced, "are to prevent AI from evolving and, barring that, to delay it. This chart analyzes the suggestions you made from the last meeting; some hastened AI takeover, some had little or no impact, but a few bought us more time, but for only six months. We need more."

The team desperately shouted out suggestions for the next two hours, but they had little effect on the timeline. Slowly, one by one, the team came to understand what Moto had surmised years ago and what Branco had come to understand just days ago.

"Chops, can I say a word?" Moto asked quietly. "I had hoped that by now we would have come up with the solution to the AI problem, but this is the dead end I envisioned when I started this. It's going to take a different mindset, a different paradigm than what we've been pursuing – what I've been pursuing. This team is capable of getting it done, I don't know how, but together we have to figure it out."

132

This was the first time Moto had intimated to the team that he didn't have all the answers, that he didn't have everything planned out years in advance. It was the first time he had looked vulnerable and reached out to them for help. Several team members spoke up to support Moto, but Chops used the Collar technology to block them from being heard.

Moto continued, "We need to form a team to think outside the 'Moto box' since I think that my influence is restricting our ability to reach the goal. So, Chops, please tell the team how we plan to move forward," there was a catch in Moto's voice.

Chops pulled Daisy a little closer and said, "We're forming a team of our freest-thinking people to challenge our current paradigm. Moloana, Beethoven, Lucy, Branco, and Sophie Jean will form the core of the team, pulling in resources as needed. For the next few months, they'll challenge everything we have done and plan to do. In the meantime, we'll move forward with every delaying tactic we have to buy more time."

Moto wasn't on the new team because to accomplish a paradigm shift, his influence had to be removed. When he began this effort, he had envisioned himself as coach and owner and everyone else as players, and in the beginning it seemed to work well, but he had been wrong. A team, particularly this one, is a gestalt and is greater than the sum of its parts—or its leader.

"I think of what we're doing as a game of chess," Moto explained, "we've made our opening move, established a defense, and positioned ourselves to gain an advantage. But we've been playing with no time limit - we need to start playing on the clock. New technology is an advantage today, and it will have to be our advantage in the future. The goal of the paradigm-shift team—we'll call it the PST—is to generate new technologies that could, in chess terms, allow a knight to move diagonally across the board or a pawn to move sideways. So if you have an idea, however outrageous or shitty you think it is, plop it in the litter box."

Chops had muted the participants – Moto didn't hear the giggles – and Chops interrupted. "Moto, we haven't told them what the litter box is yet. Team, we are setting up a drop box that you can access at any time. If you have a suggestion, just say 'litter box' and the Collar technology will catalogue your idea. You'll get immediate feedback, just like the brainstorming presentations, but ideas that need further investigation will go to the … P-S-T." Chops said "PST" slowly with a touch of sarcasm. Daisy giggled, and Chops fumbled to mute his Collar before the team heard her laughter.

Moto continued his soliloquy: "We'll be reorganizing our team for the next phase of the AI threat. You may get some additional assignments to do in your spare time." Chops opened everyone's speaker so Moto could hear all the laughter—some more incredulous than others. "Yeah, right – spare time – I get it," Moto continued, "I love you guys."

Hurley tried to speak but couldn't get words out. The rest of the team took up the slack, thanking Moto for their opportunities and for all he had done.

Lily's emotions about what they had all accomplished, what challenges lay ahead, her fear for Moto's health, and her love for Moto sent her out of the room to regain control in private.

Branco spoke up, "Guys, get it together. We've got work to do. Moto has done super-human things, but, shit, we have to look to the future. It's all about the future – step it up."

If the team didn't know from that statement that Branco had moved into a leadership position, Chops made it clear: "Team," Chops said a little shakily, "I've been grooming Branco for the past year to take my place. He is more than capable and ready to take over. I'll stay involved to help where I can, but I *actually* have a personal life now." The applause was tentative and uncertain—just a smattering—but the well-wishes exploded, wishing Chops the best and congratulating Branco.

When the meeting was over, Chops turned to Daisy and smiled a smile that managed to be both happy and sad.

"Don't think you're off the hook, Choppo," Daisy said. "We signed on to this for life to save the world." Chops smiled and pulled her toward him.

***

Indy awoke from a deep sleep. He felt a change coming but was unsure whether it would be good or bad. He hadn't spoken to anyone on the team since he'd gone to be with his grandmother, but he felt connected. He saw an image of Lily hugging his grandmother.

Chapter 29 – Edie and Lucy go to Europe

Lucy was to accompany Edie to Italy for the extraction and their cover was that Lucy would play the role of Edie's lover. Otis thought it was funny, but Lucy just ignored him. Edie and Lucy practiced being a couple, holding hands in a movie theatre and sleeping in the same bed in a hotel. They wanted the act to look natural.

Lucy got up to speed on the extraction plan as she would be Edie's backup. Edie had started learning Italian when she first heard that Riffy was going to Italy, and now she was almost fluent. Lucy spoke French and German like a native after spending years in Switzerland when she was in school.

Chops' received a message from Riffy—"46"— Riffy wanted to rescue forty-six people, including all the programmers, Henry, Max and Mitsy. Edie had a risk-assessment that allowed for that, so she activated the plan. Sophie Jean and Hurley identified the relatives of all the people RICKY had coerced into working for them, and they deployed teams across the world to move them to safety. The forty-four new recruits were categorized as trustworthy or not trustworthy. The trustworthy ones would be part of the team going forward; the others would be given a significant reward and placed in a comfortable location, but monitored closely.

Chops, Hurley, and Branco waited with Edie and Lucy in Edie's apartment for the car to arrive to take them to the airport, and when the taxi got there, Edie stood up and said, "Let the games begin!"

"Anything you need – we're here," Hurley said.

When they got to the cab, the role-playing began and for the next two weeks, their act couldn't stop. It was part of Edie's nature to act "big", but for Lucy, it was nerve-wracking – not relishing the center of attention. In the back of the cab, Edie cuddled up next to Lucy and said, "Thanks for coming with me. I love you."

Lucy gave Edie an adoring smile and held her hand. *Let the games begin, yeah right,* Lucy thought.

They arrived in Amsterdam, trudged through customs, and took a cab their hotel, but they were so tired that Edie's hands shook and Lucy was snapping at her for no apparent reason. They had but a few hours to sleep before meeting their Netherlands contact, so they set an alarm, fell into bed, and were soon dreaming.

Edie woke up a few minutes before the alarm was set to go off and quietly got ready, leaving a note for Lucy, and sneaked out of the room, allowing Lucy a few more minutes of much-needed rest. She scampered downstairs and onto the sidewalk, then strode a few blocks to the red-light district and found her contact's address. She slid into the alley and knocked on the back door, and when it opened, a small child beckoned her inside. Edie activated her Collar so that Chops and Hurley could listen in.

A man whose short, scraggly reddish beard and eyeglasses made him look older than his thirty years introduced himself as "Shorty". Edie responded in kind, reflecting his Dutch accent with her own, and said abruptly, "My companion is back at the hotel. Let's go over the logistics."

They reviewed Shorty's responsibilities until Edie was satisfied with his grasp of the plans. *One down, four to go*, she thought.

Returning to the hotel, Edie found Lucy awake. "I caught the meeting on my Collar," Lucy said smiling but before they could discuss it, there was a knock on the door, and Edie spun around in surprise. Lucy strolled to the door, saying, "I ordered breakfast in, sweetie." The attendant entered with a cart loaded with food, pointing to one of the platters and said, "Poffertjes with stroop, slagroom and aardbeien." They didn't need a translation, as they could see it was pancakes with syrup, cream, and strawberries. "Haring and bitterballen," he added, pointing to the herring and deep-fried meatballs. Finally, he pointed to a plate that contained the letter E made out of chocolate and said, grinning, "Edie, this is your chocoladeletter." Then he bowed, accepting Lucy's tip with a small salute, and left the room. Edie smiled at Lucy as they dove into their feast.

After checking out, they met their driver on the street. The first stop on their trek through Europe was in Belgium, near Brussels. It was a pleasant drive through the Netherlands, past fields and fields of flowers in full bloom. Edie pondered a world controlled by artificial intelligence and mourned the idea that there might be no one to appreciate the beauty, but that worry was soon replace with joy as they continued their drive through the spectacular countryside.

They arrived in Leuven before noon, even though they stopped several times along the way because of Lucy's three mugs of coffee at breakfast. The driver dropped them off at Park Michotte and said he would return in two hours. They walked to a bench surrounded by trees and flowered shrubbery, taking turns wandering around the park to stretch their legs, ensuring that one of them was always on the bench. At exactly noon, Shorty sauntered up just as Edie was just returning.

"Edie, my friend," he cried, clapping her on the shoulder as though they were the best of friends, "it is good to see you again."

"Why are you here?" Edie snapped. "We were expecting Abbie."

"Abbie is on her way but is delayed by traffic. She felt more comfortable if I was here with her." Edie squinted at him skeptically. "Ah," Shorty said with a shrug, his face wreathed in smiles, "Europe is such a small place. We all know each other."

"Just a sec," Edie said and she walked a short distance away and contacted Chops, asking if Shorty could know Abbie.

"I'm sorry," Chops replied. "I should have told you that most of our European network is connected. Only a few are lone wolves. Still I'm surprised that Shorty showed up at Abbie's meeting. I'll fill you in on all the relationships in Europe - I certainly should have done that before. I'm sorry to have put you in this position and will send Moto and Sophie Jean a message about my mistake."

"OK, Chops. I'm just so nervous," Edie said. "Thanks for confirming, we'll talk later."

Edie returned to the bench just as a 1964 robin's-egg-blue Mustang convertible pulled up and a young, pretty, energetic woman with short auburn hair and a feisty smile stepped out.

"I'm Abbie," she announced in a confident, sweet voice and the three women sat on the bench to discuss Abbie's role in the project. Shorty bowed, said adieu, and left.

Abbie's father had recently retired from the family "business" and Abbie had taken his place. Seeing Abbie on top of every detail, Edie was confident she could count on her. When Abbie stood to leave, she said with a wink, "Don't worry, girls. I won't tell anyone you are not really a couple. It takes one to know one, you know." Edie and Lucy laughed weakly as Abbie strolled back to her car.

Once Abbie was out of earshot, Lucy said, "I think we can trust her." Edie nodded.

Edie and Lucy's car picked them up to drive them to Le Chatele, France, and the doomsday site outside Paris. Neither Lucy nor Edie had visited a doomsday sight, though they had been briefed on them. They stopped at a picturesque farmhouse on a dead-end road. A hundred and eighty-six acres surrounded it with burgeoning fields of vegetables, grapes, and hemp.

"We can make baskets out of the hemp," Lucy said with a twinkle, and Edie laughed. When the driver left them with their luggage, a tall, slim man with a broad, sexy smile came out to greet them. He introduced himself as Buddy in a slow Texas drawl and took their luggage. "Ah don't get many visitors here," he said over his shoulder, carrying the heavy bags as though they were filled with cotton balls. Edie and Lucy tried not to notice this tall hunk's strong back and long black hair.

"I thought y'all might be hungry," Buddy continued, "so I put together a little supper for ya'. I know you're probably pretty tired from your travels, but I also figured y'all might want to relax a bit."

Edie mumbled something unintelligible, so Lucy piped up with "Thanks, Buddy, it's been a spell since we ate."

Edie stifled a laugh and whispered, "A spell?"

The inside of the farmhouse was cool and cozy, and the table in the large dining room was heaped with fruits, vegetables, and breads. For the next couple of hours, Edie, Lucy, and Buddy entertained each other as they discussed the extraction plans. Finally, Lucy observed, "I understand this is a doomsday site, but it seems kind of small."

Buddy cracked a grin and scraped his chair back from the table. "Let me give ya a tour." He led them to a closet door. "Go on, open it," he chided. Edie grasped the doorknob, but it wouldn't open and Lucy tried it as well, to no avail. Then Edie noticed a dot on the door about halfway up, put her finger on it, and in a few seconds the door opened into an elevator. "Nice figurin'," Buddy said as he and Lucy joined Edie in the elevator. "Down, please."

When the elevator stopped, it opened into a small room. "Is this it?" Lucy asked, but before Buddy could answer, another set of double doors opened.

"Facial recognition," Buddy explained.

Past the double doors was a vast underground area with sections for eating and meetings, and hallways branching off in every direction. "We can support 115 people at this facility,"

Buddy said matter-of-factly. "Let's take a look at the living quarters."

Edie and Lucy followed him down a corridor as if in a trance, looking left and right, up and down in amazement. When they got to the end, Buddy opened the door to a 20 x 20 room with a comfortable-looking bed, a desk, a bathroom, and a small lounge area. "This must be Moto's room," Edie guessed.

"Nope, they're all the same."

"You're kidding. There are rooms like this for 115 people?" Edie asked. "They must be set up for couples."

"Nope," Buddy repeated. "We don't have any rooms for couples. Not that we don't like couples, but sometimes relationships don't last." Clearly, this was Sophie Jean's practical, calculating influence.

Lucy said, "I prefer some privacy for … well …." She trailed off and made brief eye contact with Buddy, quickly looking away. *This guy is gorgeous, oh my God!*

After the tour, Edie and Lucy, yawning, found their way to their quarters, and went to sleep.

In the morning, they were awakened by a sun clock, which slowly lit up their room like a rising sun. There wasn't time for breakfast, but Buddy had packed a basket of goodies that made the seven-hour drive through the countryside to Zurich seem like a picnic.

They checked into a quaint hotel in the outskirts of Zurich in the middle of the afternoon and had an hour before their next contact, so they walked around, scampering hand-in-hand through a meadow filled with flowers singing "The hills are alive …." Lucy laughed as if she were in dog heaven and whined to stay when Edie said it was time to get back.

They went straight to the corner table at the hotel bar and had barely sat down when a tiny one-armed man with curly hair approached the table and helped himself to a seat.

"I'm Jackson," he said, wheezing. Edie looked concerned, so he added, "It's just my asthma. It acts up when I get excited, and I am definitely excited."

"I'm Edie and—" Edie began.

"Yes, Shorty told me all about you," Jackson interrupted.

Edie made a mental note to discuss Shorty with Chops again, as Shorty's apparent tendency to contact everyone he knew in Europe could risk the project. She turned her Collar on so Chops and Hurley could listen in.

Jackson ordered wine and they talked about everything but the plan—the weather, food, movies—before they finally got to the extraction plan. Jackson was thorough, and they discussed contingencies until Edie was satisfied that Switzerland was also in good hands. Edie emphasized to Jackson that he was to take direction only from her and Lucy, not Shorty, since he was not privy to the whole plan. "Jackson, the less you know about the other plans and the less they know about yours, the safer it is for you and everyone else," Edie stressed, her eyes boring. Repeating herself incessantly, Lucy put her hand on Edie's letting her know that Jackson had gotten the message.

Edie and Lucy returned to their room and called Chops to discuss their "Shorty" concerns and how the plans were not compartmentalized as Edie had expected. Her voice trembled as she listed her concerns.

Chops and Hurley listened patiently. When she finally took a breath, Chops jumped in, his voice even, "Don't worry, Edie. Hurley and I have investigated these people thoroughly. If you analyze the threat that one person that our research says is trustworthy will cause issues solely because that person is in contact with other members of the team, you'll find the risk negligible. If Shorty or any other member of the team hadn't been a hundred percent vetted, we might have had an issue, but Hurley and I will stake our lives on the team we have in Europe."

"On my life," Hurley added.

Edie was quiet. Then she said, "Well, I hope I didn't mess anything up by telling Abbie and Jackson not to talk to anyone else, especially Shorty."

"We're already mitigating that, Edie," Chops said. "No problem. Just continue with your plan."

"Okay," Edie said exhaling in a full whisper.

After the call, Lucy put her arm around Edie's shoulder and said softly, "Riffy will be okay, Edie. Don't worry – I'll make sure of it."

They talked for hours, sitting cross-legged on the floor, and when they finally went to bed, Lucy fell asleep quickly, but for Edie it took a staggering thirty seconds longer.

Moto went to his appointment with the doctor, expecting that the second test result would be negative and his ALT liver enzyme would be back to normal. As he waited in the doctor's office, he thought about the risks of using the electrical grid for their communications, as it might eventually be compromised, so he considered an alternative medium—water.

The doctor, a young slim man wearing an oversized frock, walked in with a chubby Hispanic technician.

"Ultrasound?" Moto whined, and the doctor nodded.

It took forty-five minutes to complete, but a minute or two into the ultrasound, Moto realized that he could work on his new technology idea while he lay still. When the forty-five minutes was up, Moto was deep in thought and a little disappointed that the ultrasound was over.

"So what are the results?" he asked the perky young technician.

"They'll be available in three days," she replied.

"Three days?" Moto's face fell.

"Well, I shouldn't say anything, but your liver looks great, so don't worry too much."

Moto strolled to a park across the street to think about using water as a medium for their quantum network. When he called Lily a couple of hours later, he told her what the technician had said.

"That's encouraging, don't you think?" she replied.

"I suppose," Moto said, occupied with his thoughts.

"I'm coming over. Okay?"

Moto managed a "Love to see you."

* * *

Hurley had hacked into the ultrasound machine at the doctor's office, and Sophie Jean, Lily, and a hand-picked technician had watched it in real-time, copied it to a hard drive, and sent it to two radiologists and within an hour they had the results confirmed. Sophie Jean did not want to tell Moto that they already had the results, as she wanted him to have confidence in his own doctor, so when Moto called Lily, Sophie Jean motioned to her not to give him the results yet. "Moto can't know we're monitoring this," Sophie Jean whispered, "he's probably been working on a new tech idea."

Hurly got up and left to go back to his office but came back, mumbling something about this being *his* office. Sophie Jean and Lily smiled, giggling as they walked out the door, content that the test results had removed the most serious illnesses from consideration.

When Moto got home, he took a shower. As the water ran down his body and pooled on the tile, it rippled, moving in patterns. His mind raced, searching for a way to build a quantum computer out of water as he had done with electricity, but he struggled with the need for more power than water could provide. Lost in thought, he almost didn't notice the shower curtain open as Lily stepped inside. They embraced, but Moto was so wrapped up in his thoughts that he couldn't stop himself from telling Lily what he had been working on and his struggles to find a way to make it work. Lily caressed the back of his neck under his hair and whispered, "It's not the same."

"What did you say?" Moto's mind newly afire.

Lily pulled back from the embrace so she could look at his face. "I said it's not the same."

Moto's mind exploded with new possibilities as he realized he could remove the constraints and methodology of his electrical grid quantum computer. His mind became excited, along with the rest of his body. He pulled Lily closer and suddenly, in the height of passion Moto yelled "Eureka!"

Lily laughed, first a giggle and then harder and harder until she could barely stand, and the two of them slid to the floor of the shower in a heap, embracing and laughing and kissing.

"I should tell Sophie Jean that our love-making helped free my mind for new technology ideas," Moto said excitedly.

"Let's just keep this one to ourselves," Lily cooed.

They sat together on the living room loveseat, Lily with her legs folded beneath her and Moto leaning back comfortably, his robe lying next to him. "I wish Indy were back," he mused. "He would be a big help with this new tech."

The doorbell rang and Moto glanced at the monitor and jumped as if he'd been shot. He leapt to the door and cried "Indy!" before he'd even opened it.

Indy stood with a surprised look on his face as Moto jumped onto him, hugging him. "I've never hugged a naked guy before," Indy said, smiling, beaming at Lily.

147

Lily brought Moto's robe to him as Indy stepped inside. It was only then that Moto noticed a large older woman behind Indy.

Bohdi snickered as she introduced herself as Indy's grandmother, keeping her eyes focused above Moto's waist until he quickly grabbed the robe from Lily and covered himself. She was kind-looking but she wore socks on the outside of her shoes and on her hands.

Lily beckoned her in and ushered her to a comfortable chair.

"I am not long for this world," Bohdi said without preamble. "I have even less time than Indy understands. But that doesn't mean I will be leaving. My grandson," Bohdi looked at Indy and smiled as only a grandmother can, "has special abilities, as do I." She leaned forward and whispered, "I can see that you two have a bond that goes beyond the wakened world." Lily and Moto clutched each other's hands, hanging on each word. She continued, "When I am gone, I will help you meet the challenges you have before you, as we can see more from the other realm and understand better the consequences of our actions. Through Indy, I will guide you to help keep the world safe."

Bodhi touched their souls and Moto reached to touch Bohdi's hand, but she leaned her head back in her chair and turned toward Indy, who had kneeled beside her. "Thank you, Indy, dear," she murmured. "I chose this time. Don't grieve. I'll always be nearby."

She closed her eyes and took a last, deep breath. Her presence filled the room. Indy, Lily, and Moto sat in silence, unable to move and then Indy bent over his beloved grandmother and kissed her forehead. He picked her up in his arms and walked out the door without a look back.

Moto thought about his grandfather, Eli, and his emotions rose like a great tide, filling him but then leaving him empty. He fell deep asleep, splayed across the couch. Lily crept into the bedroom and lay down, exhausted.

***

The doorbell rang at 2:00 the next afternoon. Still on the couch, Moto opened his eyes, squinting at the afternoon light.

"It's Indy," Moto said to Lily as she came out from the bedroom and then he closed his eyes. Lily opened the door and gave Indy a long hug and her condolences.

Indy strode to the chair Bohdi had passed away in the night before and sat down. "I wanted to thank you for your kindness last night," he said, his voice tired, "and to tell you that I sent Bohdi's body back to Australia to be with the rest of our family."

Moto's eyes remained closed. "She came to me in my dreams last night," he said, "You were there. So was Lily."

"She was in mine, too," Lily said, her voice small.

Moto sighed and sat up. "We were walking down a dirt road and came to a fork. We went left. Then we came to another fork and we went right. We did this several times until we came to a fork with three choices, and we backtracked and started over. We did this four times, each time making different choices, but every time we ended up at the fork in the road with three choices." Moto looked first at Indy, then at Lily.

149

"What did you dream last night, Lily?" Indy asked.

Lily blushed. "It's silly."

"No," Indy assured her. "Just tell us what you remember."

"Well, I was Dorothy, Moto was the Cowardly Lion, your grandmother was the Scarecrow, and you were the Tin Man. We were walking down the yellow brick road, but it came to an end and we didn't know what to do. That was it."

Indy nodded. "My dream had different imagery but the same message that we are going down a path, and the time for making the critical choice is soon to come. My grandmother will help us when it's time."

They sat in silence, each in their own thoughts. Lily left to get dressed, and Moto and Indy talked quietly around the edges of Moto's new tech idea, launching into an animated discussion about water. Lily glanced at them as she walked out the door, unnoticed.

It didn't take long for Indy to come up to speed on the requirements of Moto's new H_2O tech, as he called it - Indy understood the electrical grid network better than anyone. Moto had been trying to retrofit his electricity-based quantum computer and network to a water-based medium, but it wasn't until Lily expanded his paradigm that he could think freely. Indy was quick to see the pitfalls of retrofitting, so they addressed their design from the ground up, beginning with an examination of how water and electricity differ. The quantum computer running a virtual processor inside an electrical grid was pretty simple compared to using water as a medium. Their challenges were how to generate power, where and how to store information, and how to stabilize the fluid environment. It was like their computer had broken apart and was floating away in a million pieces, and they had to find a way to put it back together.

Lily returned to check on Moto the next morning and he and Indy were still in deep discussion. Lily licked Moto lightly on the nose, ordered some food and made them get up, wash, and sit at the table to eat.

She asked them to explain at a twelfth-grade level the technology issues they were having, as though they were speaking to someone who had never been a programmer. They struggled at first, but were able to elaborate on the power and stabilization issues. Lily held up her hand. "The amount of power you will be able to generate," she said, "will drive everything you can accomplish."

Indy and Moto exchanged surprised looks. She was right: They were spending a lot of time discussing stabilization without knowing how much power they could generate. They went back to their discussion of power generation, but Lily saw that they were soon outside their depth.

"Do we know anyone who has been working on alternate methods of generating power?" she asked innocently.

"You're right," Moto said. "This isn't our thing. I have a cousin—"

"I'll have Hurley make the contact," Lily interrupted. "Continue with your design, assuming unlimited power."

"And so now you know why I love her so much," Moto said to Indy.

"But she didn't even ask who your cousin is," Indy said, raising his eyebrows.

Lily winked at Indy. "I'll be back tomorrow," she said, pulling Moto up by one arm. "You boys can talk at the gym while you are working out just as easily as you can on the couch."

Indy shook his head at Moto as if to jokingly question his manhood, but Moto confidently blew Lily a kiss and she tossed one back as she dashed out the door.

Stumbling down to the gym, Moto got on his treadmill, while Indy eyed what looked to be a new rowing machine.

"I didn't used to have a rowing machine," Moto said, and then he remembered lamenting to Lily that he didn't have one. *Thanks, Lily.*

***

Hurley didn't have to search for Moto's cousin. Daisy Sue had been born in Madagascar, raised in California, and now lived on Maui. She was a chemical engineer, but didn't have a very successful career, as she'd been fired from every job because she couldn't conform to her employer's requirements. She would present off-the-wall ideas that were of no interest to her money-grubbing employers and she had written a paper on chemical fusion that the scientific community panned.

Moto directed that Daisy Sue (and several of his relatives) receive a salary from Ball and Chain for "research", but she hadn't provided the company with reports or any information on what she had been working on. Hurley sent Otis to Maui to talk to her, as Otis had a double major in chemical engineering and computer science. Lily was surprised to hear about Otis' background, but given her dreams, Bohdi, and Indy, she was resigned to what she called coincidental destiny.

Chapter 31 – Edie and Lucy Keep Trekkin'

Lucy let Edie sleep in. It was two hours past the time they were supposed to leave for Munich, but Lucy had secretly arranged to meet three hours later, ordered breakfast, and packed their suitcases.

"Lucy! We're late!" Edie yelled suddenly, awake at last.

"Everything's under control, Edie. You needed some sleep." Lucy patted her on the head. "I pushed Munich back three hours."

"Oh, you sweetie," Edie exclaimed, rolling back over.

"Have some breakfast and get cleaned up. Vacation's over. Our car will be here in an hour."

The four-hour drive to Munich, or "munchkin," as Edie called it was relaxing with no talk of the extraction plan. They drove around a large lake, and when they edged into Lichtenstein, Edie joked that, when they got to the Bierhaus in Germany, she would "lick-them-steins." Lucy rolled her eyes.

$$***$$

Brandy was irked at having the meeting pushed off a few hours, but the money was good, so he kept his feelings hidden. He was a brawny German with a large head and a big, toothy smile. He had lived in Munich his whole life, and his friends called him "the Protector" because of his fierce loyalty. Brandy looked intimidating and he was, but he had a big heart. Chops had met with Brandy personally a few weeks before and they had hit it off. He was determined to make Chops proud by protecting the people to be rescued from RICKY.

Lucy and Edie climbed out of the car in front of the Eggroll Inn, an inviting restaurant located on a busy street in Munich and, eager to please, Brandy dashed toward them.

Lucy leapt backward and yelled, "Hey!"

Brandy stopped and put his hands up. "Sorry," he said with a broad smile. "I didn't mean to scare you. I'm Brandy."

"Yeah, big, strange guys like you running at us are a little intimidating," Lucy replied smiling with a wink. Then she gave him a hug and Edie followed suit.

Brandy felt like he was hosting a family reunion. "Danke," he said with a little bow. "I'll keep it in English for now, but later in the Haus we'll be singing German drinking songs." Edie smiled broadly, but Lucy gritted her teeth, trying to remain in control.

Brandy ushered them inside, gesturing broadly, "Wilkommen!", he shouted, brandishing menus. "Mein establishment makes the best Filipino cuisine in Germany." A server brought in a big basket of lumpia and a bowl of sauce and asked what they wanted to drink.

"Bier, bitte!" Edie called out.

"Just water," Lucy said, nibbling at one of the eggrolls, but after her second nibble, she jerked her head up and said, "I'll have a beer, too."

Their drinks arrived quickly. Lifting her stein toward Brandy, Edie began, "To Riffy, the man who—"

Edie stopped abruptly, and Lucy looked over, her stein still in the air and Brandy put his hand on Edie's.

"Don't worry, Edie," he whispered gruffly, his face close to hers. Chops told me about you and Riffy and how hard all this is for you, but this plan will work. We'll make it work. You'll get him back, safe and sound."

Edie sighed, and lifted her stein higher. "Yes, to Riffy and his safe return," she announced, followed by a lengthy quaff.

Lucy looked at Brandy admiringly, a look that could have been interpreted as respect or lust. "Don't go there, Lucy," Brandy said with a wink and she blushed, offering a shy smile.

Oblivious, Edie dove into the eggrolls. "Don't worry, Lucy," Edie said without looking up, "There's no chicken in the eggrolls. These are shrimp and pork." Lucy shook her head with a smile, thankful that Edie's chicken allergy had been considered.

"Our best eggrolls are made of turkey," Brandy said, waving a waiter over, "but I wasn't sure that Edie was okay with that. I'll get some for you, Lucy."

After eating an incomprehensible number of eggrolls, they talked about the extraction plan. Before long, Edie and Lucy were comfortable with Brandy's understanding and commitment. They clarified a few small details, but Brandy was on top of everything.

A few hours later, Brandy shuttled Edie and Lucy off to bed in rooms he'd prepared at his inn. "Breakfast meeting at oh-seven-hundred," he called as they walked up the stairs. Lucy turned and smiled sweetly.

Once out of earshot, Edie yawned, looking more relaxed than she had been on the entire trip and asked, "How many eggrolls did I eat, Lucy?".

"You had fourteen and a beer and a half," Lucy replied. "I had seventeen eggrolls and two beers."

"Ah, Lucy, I didn't need the beer" Edie said as she stretched out on the bed. "Brandy made me feel better."

Lucy lay back and thought about Otis.

***

The wake-up call came at 6:30 the next morning, but Edie and Lucy were already up and packed. Laughing at a shared joke, they tumbled down for breakfast to cover a couple details of the extraction plan. Brandy was there with a platter of meats, cheeses, eggs, and pastries. A moment before Lucy reached for a pastry, Brandy asked what they wanted to eat, and she realized that all the food on that large plate was just for Brandy. Before they could answer, Brandy shouted, "Two more for the ladies!" The waiter ran back to the kitchen and quickly brought out two more platters, just as large. "Now, where were we?" Brandy said, waving the waiter away.

Edie was grateful that Chops and Hurley had found this big lug. When they finished, Edie and Lucy's platters were almost empty.

They said their goodbyes and walked towards their ride, but Edie stopped and turned saying, "Hey, Brandy," and when he smiled, she winked and nodded.

Once out of sight, Brandy hobbled to a bench and sat, wincing. The adrenalin of the visit had let him withstand the agonizing pain from his arthritis so that Edie and Lucy would have the highest confidence in him.

"He's quite a man," Edie said to Lucy once they were in the car, but Lucy didn't look up. *It's going to be a long drive to Innsbruck*, Edie thought.

Lucy was upset with herself for being attracted to Brandy. After all, she loved Otis. Then it occurred to her that if Otis had met Brandy, he might also be attracted to him, and that thought bothered her even more. She longed be back home with Otis and just leaned against the window, gazing at the passing scenery.

Edie put her arm around Lucy's shoulder and they rode silently for the couple of hours it took to get to Innsbruck.

When they arrived, the doorman handed Edie and Lucy their keys and took their luggage. Edie suggested they go for a walk to stretch their legs since they had a couple of hours before their next contact.

"Good idea," Lucy said, grateful for the distraction.

They strolled down one street after another in silence until Edie finally asked, "Is there anything you want to talk about?"

"I want to talk to Otis. I don't think I can last much longer playing these stupid roles."

Edie nodded, sat Lucy down on a nearby bench, and walked away saying, "Go for it, Lucy. Collar him."

So she did, even though she knew it was really early in the morning. "Hey, baby," Otis said, his voice gruff and sweet at the same time, "Que pasa?"

Lucy already felt better seeing his face and shirtless chest against his rumpled sheets.

"I miss you, babe," Lucy murmured. "I just called to hear your voice."

"Wish you were here, Lucy," Otis said, yawning.

"Me, too. Gotta go," Lucy cooed.

The call lasted nine seconds, but Lucy felt as refreshed as if she'd slept eight hours. She beckoned Edie to come back, and when Edie sat on the bench with her, Lucy said, "Let's make sure Riffy gets back safely."

"Ain't love grand?" Edie said, giving Lucy's arm a squeeze. Lucy nodded, wiped away a tear that was threatening to escape the outside corner of her eye, and took Edie's arm. Chatting and laughing, they walked briskly back to the hotel.

Moto hated the water. He became anxious just getting near it and just didn't even like getting wet. He'd had his bathtub removed from his apartment and had designed a "dry shower" to remove dirt and odors from his body without liquid and considered marketing it but was too embarrassed to admit his aversion. But now that his next big technological breakthrough was to use water as a communication medium, he wasn't going to be able to avoid it and decided to hire a swim coach to help him overcome his anxieties.

Tilly arrived at Moto's apartment at about 6:00, along with Lily and someone Moto didn't know–Griffin. Moto was surprised at this hearty cast but shrugged it off, understanding that he really didn't have any privacy. Lily hugged Moto and introduced Tilly and Griffin – Tilly the swim coach and Griffin her boyfriend.

Tilly kept her head down and said matter-of-factly, "The car will pick us up in eighteen minutes." Moto cocked his head.

"We're going back to Maui," Lily said quietly.

Moto turned to Lily saying, "Thank Sophie Jean for me." And she nodded.

Griffin stood in the middle of the room. He had strange skin that made him look from a distance like he had some kind of serious mottling disease, but up close his skin looked smooth and beautiful. "I'm Tilly's boyfriend and just along for the ride because Tilly and I love the beach," Griffin said. "You can trust me though. As you might expect, I've had some big-time grilling from Hurley and Chops."

Moto smiled and said, "They're pretty thorough. Good to meet you, Griffin. Enjoy the ride."

When the car arrived, Lily went to Moto's bedroom and retrieved the suitcase she had secretly packed for him, while Hurley loaded everyone's luggage into the limo. Moto saluted Hurley as they all got into the car, and Hurley responded with a sweeping bow as they took off.

Hurley followed Moto's car, but his tech saw traffic issues ahead, so he pulled in front of them and collared Lily to have the driver follow him. He put on his flashing blues and motored through the city, avoiding the bad traffic and opening lanes.

They arrived at the airport twenty minutes before the flight was to leave and even the TSA pre-check line moved slowly. They got to the gate only ten minutes before the flight was to leave, so the door to the jet way had already closed. Hurley flashed some kind of official badge he had forged, and the door opened. Hurley went in first and had all the passengers who had been upgraded to first class moved back to coach, which caused considerable grumbling. When Moto and the rest boarded, they weren't aware of the commotion they had caused, so they just sat down in their comfortable seats and waited quietly for the flight to Maui to begin.

The flight attendants gave them the VIP treatment while the other first class passengers shook their heads, whispering to each other. Noticing their angst, Lily leaned over to one of the other passengers and said, "We're here to save the world," eliciting a laugh. For the rest of the flight the other passengers were pleasant and friendly. *If they only knew*, Lily thought.

***

Daisy Sue lived in Kula, which as the crow flies is only a few miles down Haleakalā to Wailea, where Moto and the gang were staying. However, the drive to Wailea on public roads takes an hour as they go north to Kahului before they switch south back to Wailea. Many celebrities live on Maui, but a notable one owns a large parcel of land that includes a private road that cuts the travel time to seven minutes. Daisy Sue had befriended this celebrity, to whom she referred to as "O", and received permission to use the private short-cut during Moto's stay.

Otis had preceded them to Maui and met them at the airport. "We're going straight to Daisy Sue's for a late lunch," he told them as they scrambled into the car. "After that, we'll head down to your hotel." Otis drove up Haleakala Highway toward Kula as the team gazed out the windows at the beautiful day that spread across the island.

When they arrived at Daisy Sue's, Moto went in first and saw Daisy Sue sitting in a wheelchair with a tall back support. His jaw dropped and he hugged her gently.

Daisy Sue said, "Kind of a shock, huh, Mo?"

"Not what I expected, cuz," he replied, shaking his head.

"I've had some back issues lately, but it's getting better. I'm hoping to get out of this highchair in the next few weeks," she said, smiling warmly at the cousin she hadn't seen in years.

"I told you skydiving was dangerous." Moto quipped; Daisy Sue shook her head with a smile.

They were already deep in discussion about the H_2O project when Otis and the rest of the team joined them inside, Otis asking, "Am I too early?"

"No, your timing is perfect," she replied with a laugh.

After introductions, they enjoyed a long, leisurely lunch before Otis took the team to their hotel, stopping several times to open security gates. Once inside the hotel, Moto and Lily said good night to Tilly and Griffin. The schedule for the week was swim lessons for Moto in the morning, when the water was clear and calm, and technical discussions with Daisy Sue and Otis in the afternoon, when it was "coolah" up the mountain in Kula (Koo-lah).

Griffin looked forward to the afternoons when he and Tilly could enjoy the beautiful resort – right on the Pacific Ocean with a spectacular view of Maui's gorgeous sunsets. He draped his arm around her as they walked to their room; where Tilly sat to review her notes for Moto's training. She was so deep in thought that when Griffin asked if she wanted to go downstairs to see the sunset, she didn't even look up. With a sigh, he wrote her a note about where he was going and taped it to her phone. A couple of hours later, he returned with dinner just as Tilly looked up from her notes.

"Mm looks good," Tilly said, digging into the food and Griffin smiled.

***

Lily, Griffin, Moto, and Tilly met for breakfast as the restaurant opened. "There are no dietary restrictions today, Moto," Tilly announced. "You won't be in water deep enough to cramp."

Tilly walked to the buffet and returned with a plate filled with scrambled eggs and nothing else and ate by making small, intricate trenches in her breakfast. When she had finished half of her meal, she announced she was going down to the beach to prepare for the swim class.

After she had left, Griffin said, "Tilly's the best swim instructor in the world." Moto and Lily looked sideways at Tilly's half-eaten, trenched plate of eggs and nodded politely.

When they got to the beach, Tilly was standing next to some trenches she had dug in the sand, similar to the pattern she had made in her scrambled eggs. She motioned for Moto to stand next to her, and speaking in a slightly North London accent, she told Griffin and Lily to go somewhere else and enjoy the morning. She directed Moto to stand at one end of the trench and to remember the last time he had been in the water. He had a clear and frightening memory of it.

"Now look at the sand," Tilly pointed. Moto looked down and then back up at Tilly. "Please look at the pattern in the sand until I tell you to look up," Tilly said, her accent now more pronounced and bordering on Irish. Moto stared at the sand for almost a minute. "Now close your eyes and picture the patterns in your mind," Tilly instructed in a full on Scottish brogue.

Moto closed his eyes and could picture the pattern clearly. Her Scottish accent stuck in his head. Moto was unsure this would work, but was not annoyed; rather, he felt content. "Now open your eyes and look at me, Moto," Tilly said in Gaelic. Moto understood what she said, though he didn't speak Gaelic, and turned to her. He saw the pattern on her face, making her look like an angel. "Look at the water, Moto," Tilly said. Moto turned to the ocean and could still see the pattern.

He walked to the water and went in up to his waist, feeling calm as the gentle waves lapped at his stomach. Then he looked around and suddenly couldn't see the pattern. Feeling anxious, he ran back to the beach.

"Good first lesson, Moto!" Tilly said in her native Midwestern accent. "We'll continue tomorrow. I suggest you have a plate of scrambled eggs for breakfast."

***

Moto Collared Otis to let him know he would be ready to meet with Daisy Sue after he showered. They were back in Kula within a few minutes. Otis and Daisy Sue had been working on ideas for power generation in the water medium. Otis had enough expertise in chemical engineering to understand what Daisy Sue was trying to accomplish and had enough of a background in both mechanical and electrical engineering to provide Daisy Sue with what she needed.

"Conceptually," Daisy Sue began, "we are designing a sphere of energy that moves within water. The energy sphere is created by a chain reaction that builds from simple kinetic energy to a chemical reaction to nuclear fusion, if needed."

Otis smiled and nodded to Moto. "So you're creating nuclear fusion out of nothing?" Moto asked.

"It's a two-step process," she replied. "First we create the energy sphere with a chemical reaction. The technology I am proposing creates more energy than it uses, but we need nuclear fusion if we want unlimited power. You don't have to be a nuclear engineer to do that."

Moto saw why she hadn't done well in a corporate environment. The way she talked gave little confidence that she knew what she was doing, but where others might view her as off-beat, Moto saw passion and insight.

"The world screwed up when they didn't listen to you," Moto said. He tried to hug her, but it was hard and awkward because of the wheelchair, then Otis tried and failed and they all ended up laughing.

It grew late and Daisy Sue waived them off so Otis and Moto headed back to the hotel and when they got there, Moto kissed Otis on the lips, placing his hand gently on Otis' shoulder, and went up to his room, where Lily awaited.

Otis was proud that he had become strategic in developing this new tech, but after Moto's kiss, Otis could think only about how much he missed Lucy.

***

They met again for breakfast the next morning and Tilly filled her plate with scrambled eggs again, while Moto got fruit and breakfast meats. Tilly ate the eggs by making intricate trenches again, but in a different pattern. Moto watched her and arranged his food similarly, comparing his plate with Tilly's before he began to eat. Tilly stood up abruptly and said she would meet Moto at the beach when he was ready.

Lily leaned over to Moto and whispered, "I drew a very similar pattern in a dream last night, but it wasn't exact."

"We'll have to go see Sophie Jean again when we're done here," Moto said smiling. Holding out a cloth napkin, he asked, "Can you draw the pattern for me?"

She showed Moto how it was different by moving his breakfast ham and honeydew melon around.

"Can I have that?" Griffin asked, pointing to Moto's ham and Moto smiled, pushing the plate toward him.

After breakfast, Lily and Griffin went on a long beach walk while Moto joined Tilly on the beach. She had drawn a new pattern in the sand next to the previous day's pattern. She asked Moto to stand and look at the new pattern, alternating closing his eyes and opening them so as to imprint it in his mind. Then she directed him to do the same for the previous day's pattern.

Moto felt the calm returning and he walked toward the water with Tilly close behind, waist deep, just beyond where the waves broke.

"Can you still picture yesterday's pattern?" Tilly asked and Moto nodded. "Look up at the sky and visualize the new pattern." As he followed her instructions, Tilly gently leaned him back in the water, holding him there as he gazed at the pattern in the sky. Moto felt like he was floating, confident, he turned his thoughts to Lily's dream pattern and wondered if visualizing that would be even better, but when he pictured it, he became anxious and flailed in the water, fighting Tilly's gentle grasp.

"Visualize the first pattern," Tilly said calmly as she worked to hold him in position. Once he had calmed down, Tilly guided him to his feet and asked what had happened. Moto told her about Lily's dream, her pattern, and what he had done while floating.

After a few thoughtful seconds, Tilly said sharply, "That was Lily's pattern. Please don't do this again. We are on a timetable to complete your training and we can't be distracted like this. Today's lesson is done." Moto watched sadly as she strode to shore and out of sight.

Chapter 33 – Edie and Lucy Get to Italy

When Edie and Lucy returned to their hotel in Innsbruck, a young woman with strawberry-blond hair was waiting for them. She broke out in a big smile as they walked up. "Hi, my name is Ginger," she said, "but you can call me Ginny."

Thank God there's not another Ginger, Lucy thought. The three spent the evening talking and laughing and reviewing the extraction plan and Edie grew comfortable. Although Ginny was very young and fun loving, she had the details of the plan down cold.

When Edie and Lucy returned to their room, they received an urgent Collar request from Moloana.

"Hello, ladies," Moloana began, "everything is fine, but I'm in Assisi to take over the extraction troops. The local contact is in the hospital - she broke both her legs in a skiing accident but she's okay. I'm completely up to speed and have already spoken to her and her contacts at length. I'm here waiting for you," Moloana said cheerfully.

Edie and Lucy were shocked but recovered quickly and managed "OK, Moloana, thanks. It'll be great to see you tomorrow." Edie knew and respected Moloana, an ex-Navy SEAL who had managed several SEAL operations. Lucy didn't know her well, but if Edie was happy, she was fine with it. They crawled into bed, a seven-hour drive ahead of them the next morning. "Are you sure you're okay with this?" Lucy asked as she turned out the light, knowing Edie's uncomfortableness with change.

"I think so. I mean Moloana is great and I trust her – in fact more so than a stranger we hadn't met yet," Edie said, but she had a nagging fear that she couldn't get rid of. She closed her eyes and opened them only when she heard the morning's wake-up call.

Edie and Lucy were grumpy. There was no hot water in the shower, and the coffee they had with breakfast was cold. But after a couple of hours of beautiful scenery and a nice lunch on the road, they lightened up and had a pleasant drive.

As they drew close to Assisi, Lucy reflected on her earlier bad mood. "Sorry I was in such a snit this morning," she said to Edie. "We're both under a lot of pressure to make sure this extraction is successful."

Edie nodded, "Yeah, me too – time to get focused." As they pulled up to the Palazzo Gattini hotel, Edie whispered, "Let the games begin."

The plan was "simple." Steal RICKY's AI code, destroy all backups, destroy the RICKY organization, and get everyone out alive to a safe place. *Yeah, simple.* Riffy had given Hurley enough information about the organization to identify the top echelon of the RICKY bosses and map their daily routines. The Palazzo was one of their favorite indulgences, so Hurley had placed operatives at the hotel.

First up was to plant communication devices on the bosses. Chops had arranged for a network of eavesdroppers to track and report private conversations using new tech monitoring implants: the Nail, the Worm, and the Polyp. The Nail was applied by a manicurist on the toe or fingernail, lasting only few weeks since the finger and toenails grew out. The Worm was a device that when placed in the ear, would work its way into the eardrum. In addition to monitoring, it could transmit sounds and voices to the host, sounding like quiet whispers. And the third implant was the Polyp, which when ingested, passed through the stomach where it attached itself to the intestinal wall releasing a chemical that would cause a polyp to grow around it. All of these implanted devices were undetectable and connected to the electrical grid network.

***

After checking in, Edie and Lucy went to the bar to meet with Moloana. The plan had been for Edie to go in first and flirt with Moloana like she was picking her up and then Lucy would come in, act jealous, and make a scene. But Edie felt this role-playing now detracted from their mission, so they walked into the bar together. After a drink with Moloana, they strolled across the street to a park to discuss the plan. Moloana was ready, answering everything that Edie and Lucy posed – giving even more details than they wanted to hear. Lucy handed her the tracking implants for the operatives working at the Palazzo. In the morning, the whole team would be able to monitor RICKY's bosses.

Hurley gave the bosses code names. The top two in the Mexican cartel, code-named Big Enchilada and Hot Taco, came to the Palazzo every week for lunch, a massage, and a manicure. They would be implanted with all three tracking devices to determine how effective they were in the field before implanting the rest of the bosses.

Big Enchilada and Hot Taco arrived a little before 11:00 that morning and were greeted royally by the Palazzo staff and whisked into a private dining room where a table for eight was set with two places. They sat at either end of the table, and a beautiful woman with long flowing black hair, wearing a brightly colored Mexican vestida came in with two plates. "Nacho Especiale," she said as she set the food down, and on each plate were three nachos, each prepared differently. "Uno de mis favoritos," she said, pointing to the nacho with chorizo and guacamole.

The bosses gulped it down and then finished the rest of the plate. "Delicioso!" Hot Taco declared.

The woman cleared the plates and returned with her hair up in a tight bun wearing a dress in the colors of the Italian flag. She carried two plates of eggplant parmigiana, shrimp scampi, and fried ravioli. They were small portions, and the bosses finished them quickly. After clearing, she came back in long flowing gold and white dress adorned with Egyptian symbols and on her head was Cleopatra's headdress atop short straight black hair. Her eyes, heavy with mascara, drew their attention as she served two small dessert plates of rice pudding and baklava.

The bosses were whisked away to a private cabana near the pool, where they lay on the loungers awaiting their next indulgences. The monitoring group verified that the polyp implants were functioning as they oozed their way through the bosses' digestive systems.

The manicurist was a sleight-of-hand magician who would secretly place the Nail implant on the thumbnail before applying polish. Half the size of the thumbnail, the Nail was activated by the polish to detect the color of the skin beneath it and change its color to match, releasing a quick-drying chemical that covered the existing nail. As the host's nail grew out and was clipped and filed, the portion of the nail that held the electronics would sense when it was close to being compromised and would self-destruct, causing a meltdown of the electronics that would be undetectable to the host.

The masseuse implanted a Worm in each of the bosses' ears while performing a sensual ear and neck massage and the implant crawled through the auditory canal, attaching to the eardrum.

Buddy oversaw the monitoring team located at the Paris doomsday site. They listened in for the next two days, tweaking the implants for optimum reception, recording every aspect of the bosses' lives, from sleeping to pooping to sex to eating to angry confrontations. Buddy was surprised at the clarity of the signal and even though Big Enchilada had irritable bowel syndrome (IBD), the implant successfully attached itself to the small intestine and grew a polyp, despite the chaos in his digestive system.

Success in hand, the team implanted the devices in the Russians, the North Koreans, the Mafia, the Somalians, and a newcomer to RICKY, the neo-Nazis, providing an edited stream to Sophie Jean, whose psychoanalysis would determine tactics that Edie could use to turn the bosses against one another.

***

Sophie Jean settled into her big, white, cushy easy chair to listen to the recordings and Hurley joined her to provide an all-important sounding board. For the next two days they kept each other company—and awake and when they finished, Sophie Jean had a newfound respect and fondness for Hurley—maybe more than fondness. When Hurley put on his jacket to leave, she thanked him and lunged forward, hugging him tightly.

"Just doin' my job, ma'am," Hurley said with a smile that had its own secret meaning—not so secret to Sophie Jean.

***

Riffy and Remi were under RICKY's constant supervision, especially now that they were getting close to going into production with their AI programmer. Until now, Riffy had had some freedom because of his position in the organization, but the bosses clamped down on him, too. Guards surrounded the compound, and no one was allowed to go out, even for a sandwich.

Back in his room, Riffy received the code "BT" on his razor. He wracked his brains for a few seconds, and then realized that it stood for bread and tuna. That evening he went down to the kitchen at 9:30.

"I'm starving, dude," Riffy said to the kitchen guard. "I missed all my meals today. Can I make myself a couple of tuna sandwiches?" The kitchen guard shrugged, and Riffy went to the pantry. The newly delivered food hadn't been put away yet and he rifled through the cans of tuna and loaves of bread, making a commotion as tuna cans rolled across the floor.

"Hey!" the guard yelled, walking toward him.

"Give me a break!" Riffy yelled back. The guard continued toward him, but Riffy found the can and the loaf of bread marked with "M" before the guard reached him. Grabbing them, he snapped, "I'm going to take these up to my room."

"Whatever," the guard spat, returning to his post and Riffy dashed upstairs.

He opened the can of tuna and ripped the plastic off the end of the loaf of bread. Performing for the cameras, he scooped some tuna onto the bread and took several big bites. His stomach was in knots, and each bite was sickening. Then he saw the Collars - one in the bread and the one in the can of tuna. The Collar in the tuna was a replica of the necklace he had worn since he came to RICKY. He dropped the can of tuna "accidentally," and as he bent over to retrieve it, he pulled the necklace out of the can and hid it in his hand. Then he cursed loudly as he walked to the shower, saying, "Damn! I'm going to have to wash this smelly crap off of me." In the shower he used a washcloth to hide and clean off the Collar, switching his old necklace with the Collar while toweling off. He "eliminated" the old necklace while sitting on the toilet. He heard Edie's voice say "I love you" through the Collar, but continued his act for the cameras.

Remi knocked on Riffy's door, and he answered it wrapped in a towel. "How nice to see you, Remi," he said suggestively, "Come on in."

"Get over yourself, Riffy," she replied. "I just came back from the kitchen. They wouldn't let me in but told me that you might have some food up here. I'm starving."

"Sure," Riffy snapped, tossing the bread at her. "Take this and the rest of the tuna and leave me alone." Remi took the food and thanked him as he slammed the door in her face. She went back to her room, and after dropping the food, using her own

175

sleight of hand, a shower, and a shit, Remi had her Collar on as well.

Knowing that Riffy and Remi had Collars, gave Edie a new level of confidence about the job that Moloana was doing. Not that she doubted her, but smuggling the Collars onto the compound was slick. She couldn't help feeling joyous about this initial success, but she muted her happiness with determination - she couldn't take her eye off the ball for a moment if she was going to get Riffy, Remi, and the others out safely. And, she couldn't lose sight that this AI development could be a threat to humanity.

Chapter 34 – Moto's Breakthroughs

Moto's abruptly ended session with Tilly, although disappointing, made him determined to follow her instructions exactly going forward. Since there was nothing he could do about it until the next day, Otis picked him up early, but once he got to Daisy Sue's, his mind raced and he entered a manic state. He had always had detailed computer programming to perform when this happened to him, but now all he had to focus on was his H_2O network plans. He lay on the couch, listening and pretending to sleep, but his mind was electric—as active as ever and without coding to focus him, he had the freedom to analyze, design, and conjecture, but his thoughts were difficult to control.

"The surface of the water will be disturbed by...." Daisy Sue's voice faded as Moto saw a bubble floating in the water. The exterior of the bubble was an energy source, protecting and feeding what was inside. The bubble burst occasionally but then put itself back together again, functioning as a quantum computer connected to other quantum computers floating nearby. With each sentence he heard from Otis and Daisy Sue, Moto's mind went in a hundred directions.

***

Driving Moto back to the hotel, Otis asked, "How was your nap?"

"Good." Moto said, not wanting to distract Otis from the work he and Daisy Sue were doing.

Otis glanced at Moto. "Just 'good'?" he asked Moto probingly.

"Uh-huh."

Otis didn't press further – he trusted that Moto would share his thoughts at the right time. When they arrived at the hotel, Moto kissed Otis on the lips again, smiled, and gave him thumbs up, dashing quickly up to his room.

Otis smiled, thinking about the thumbs up from Moto as he drove back to Kula. When he got to Daisy Sue's, she was still working. "You need to get some sleep so your brain is fresh for tomorrow," he told her.

She pushed back from her desk. "You're right, Otis." This is important shit we're doing here, huh?"

"Yeah, no one else could do this," Otis said slowly as if he was just realizing their importance.

Daisy Sue nodded slowly, looking briefly into Otis's eyes.

***

When Moto walked into their room, Lily was already in bed. He leaned down to give her a kiss, and she looked up, seeing the excited smile on his face. Leaping out of bed, she grabbed a robe and put on some coffee.

"How did your afternoon go?" she asked brightly, while she curled up on the sofa.

They talked all night until Moto fell asleep just before sunrise. Lily went for a beach walk to wind down from their conversation before meeting Tilly and Griffin for breakfast.

"I don't think Moto will be down on the beach for his lesson before 9:30," Lily said as she sat down with a plate piled high from the buffet. Tilly averted her eyes and grimaced, but then looked up and nodded while Griffin stared blankly at his coffee cup.

When Lily got back to the room, Moto wasn't there, and the late night had caught up to her. *Aaaaah,* she thought and she stretched out in bed. *Griffin would have to fend for himself today.*

***

Tilly walked down the wooden planks to the beach and saw Moto standing there, smiling. Moto had dug the exact trench in the sand that she was planning to dig. "I guess we can skip the visualization portion of this morning's lesson," she said matter-of-factly. "Please keep this image in your head and walk to the water with me."

Moto nodded dutifully, but as they walked into the water, the image of the pattern in his head began to fade.

Sensing this and without even looking at him, Tilly said, "Take a deep breath, Moto. Visualize the pattern in the sand and then exhale." Moto mind wandered uncontrollably. Tilly took his hand and they waded into the waist-high surf. "Keep the pattern," Tilly whispered above the roar of the surf. "Keep the pattern." Moto breathed heavily. Tilly turned to Moto and whispered, "Take a deep breath." She drew the pattern on his forehead with her fingers, and Moto breathed deeply. "Lay back," Tilly said, her hand on the back of his neck.

Moto's mania subsided, and he was shocked and lost focus again at the thought of being able to control his manic side. Tilly sensed the change and whispered again, "Keep the pattern." When Moto had calmed down, she added, "Close your eyes, take a deep breath, and slowly breathe out through your nose."

Tilly gently pressed his forehead down, still holding his neck, but his face was under water. Then she slowly lifted him up. "Again," she said, repeating the motion, "and again."

Finally, Tilly removed her hands and said, "Again." Moto took a deep breath and dunked his head. Tilly gently lifted it back out of the water. "The session is over," she said. "Can you remember the pattern until tomorrow?

Moto nodded, knowing he would never forget the pattern that allowed him to control his mania.

Moto raced to his room to tell Lily he had a way to stop his mania at will, but found her fast asleep, so he crept out and jogged back to the beach to ponder his H_2O project. Lily came down a few hours later, and they went to happy hour to celebrate over a plate of calamari, looking out at a glowing sunset.

Moto said, "We should have done a doomsday site here."

"We still could," Lily said in all seriousness. "You know, the Big Island could be a great place for thermal energy." Moto nodded, and Lily took it as agreement that this was her next big project.

***

Moto, Lily, Griffin, and Tilly met for breakfast and as Tilly ate her omelet from right to left, Moto told her that concentrating on the pattern had allowed him to control his manic state. Tilly didn't look particularly surprised. "So do you think there's a pattern that would allow me to enter a manic state?" Moto asked.

Tilly stopped with her fork halfway to her mouth. "In my paradigm," she said, laying her fork carefully on her plate, "mania and anxiety are to be avoided. She paused and then added, almost reluctantly, "but there probably is."

"When we complete our swimming lessons, can we spend a few days looking into that?" Moto asked. Tilly nodded, already considering how to develop a pattern to allow Moto to go manic.

Tilly taught Moto to swim, first with a snorkel, then with swim goggles, then with no aids. Before he realized it, Moto looked forward to getting into the water. When the last lesson was complete, Tilly told Moto that she was ready to work with him on a pattern that would allow him to enter the manic state, but she needed a couple of days to prepare.

For the next few days, Moto spent his mornings swimming in the ocean, his afternoons in Kula with Daisy Sue and Otis, listening in on their work, and his evenings with Lily. *So this is what a vacation is like*, he thought, having never taken one before.

Lily enlisted Griffin to help her complete the reconnaissance for the Hawaii doomsday site. "Doomsday in paradise," Griffin joked, "that's a strange juxtaposition."

"We'll call it—let me think" she laughed, then turned serious, "A'I'kapu – that's Hawaiian for 'AI keep out.'"

Lily analyzed the geothermal capabilities of the islands. Because of the active volcano on the Big Island, she felt it was too unsettled to tap its power, so she settled on Maui, whose volcano, Haleakalā, was dormant, not extinct. It had last erupted in the eighteenth century, was relatively stable and had tremendous geothermal power. She settled on a remote site near Haiku, near a little-known geological hotspot. She paid cash for the property at a little over its asking price and began lining up contractors to begin construction. She wished Moloana could be there to help, but Moloana had a crucial assignment on the other side of the planet. Besides, Lily was impressed with Griffin's skills and also his rugged good looks – *nice to be around,* she thought.

Tilly called late one night to say she had the new pattern and Moto met her at the beach under a bright moonlit sky. "I have three options for you," she said, handing him a headlamp. "Are you ready?"

Moto nodded, his headlamp moving its beam of light up and down. Tilly drew the first pattern in the sand, more angular and intricate than any of the calming patterns. She motioned for Moto to gaze at the trenches she had dug and directed him to take off his clothes and go for a swim. Moto complied and waded into the water, but then Tilly called him back, she had erased the trench in the sand.

"Now picture the calming pattern," Tilly said. Moto closed his eyes, picturing the calming pattern, hearing Tilly digging in the sand. "Open your eyes, now, Moto, and focus on this new pattern," she said. Moto gazed at the pattern and felt his heart rate increase and his mind race. "Moto," Tilly said softly.

Moto continued gazing at the pattern and—bam!—he was manic. He looked up at the sky, thinking hundreds of thoughts at once.

"All you have to do to come back down is to focus on the calming pattern," Tilly said. Moto barely heard her, but he focused and his mind steadied.

Moto was ecstatic that he had found the key to mastering his bipolar disorder and he hugged Tilly tightly, forgetting he was naked. Tilly didn't mind and she clung to him before they ran into the water, splashing and jumping in the waves.

Moto floated in the surf. A lifelong weight had been lifted from his shoulders. He could control his mind now, and he could decide for himself when he needed to be Superman and when he needed to be Clark Kent. On top of the world, he compartmentalized the patterns with trigger images, picturing himself swimming to get to the calming pattern and picturing himself in front of a computer screen to get to the manic pattern. He felt free for the first time in his life.

When Moto got back to his room, Lily was working on the A'I'kapu doomsday site, but when she saw the look on his face, she took his hand and led him to the bedroom.

"Maybe we should call Sophie Jean," he said, hesitating.

"She already knows, and so do I. It can wait," Lily replied.

***

The next morning, Sophie Jean called, gushing, "I wasn't convinced it would work, but I'm glad it did." Moto grinned. "Now it's time you came home," she said quietly and signed off.

Moto sighed and reached for Lily's hand. "You know I have to stay here to work on the doomsday site," Lily said, "and I want Griffin to stay and help." She paused and squeezed his hand. "I'm attracted to him, Moto."

Moto nodded, his mind elsewhere. "I need to stay a couple days to finish up with Otis and Daisy Sue," he said. Then he looked up. "Wait—did you say you're attracted to Griffin?" Lily nodded. Moto looked down again. "I guess I'm not around enough for you," he said softly.

"It doesn't matter, Moto. We're bonded on a level way beyond the physical and emotional. I'd never do anything to hurt that," and she gazed into his eyes. Moto kissed her hand.

"I'm going to Little Beach," he announced, grabbing a towel.

Lily smiled and said, "I have a busy day today. See you tonight."

Chapter 35 – Riffy Brings Henry In

Remi was amazed at the Collar technology but felt exposed when she went to the bathroom, showered, or changed her clothes even though she was under constant surveillance by RICKY. She was brought up to date on the plan to extract everyone from the site and take RICKY down. She had been worried before, but hearing the whole plan terrified her, especially when she found out that Riffy couldn't wipe out all the backups, so it would be up to her to destroy the data at her site.

Riffy played it cool. There were three sites to destroy and he had to wipe the computers at two of the sites simultaneously while Remi handled the third. He sent a coded message asking Chops if he could trust Henry. No more than an hour later, Sophie Jean's voice on his Collar said, "Henry is 100 percent."

Riffy's Collar had been equipped with a removable section that contained a Worm, and it was up to Riffy to get the implant into Henry's ear so that the team could talk to him, explain the plan, and monitor what he was doing. Now, he just needed an excuse to stick his finger in Henry's ear....

***

Henry was always first to arrive at the Wednesday status meeting. Riffy was usually a little late, but this week he got to the conference room twenty minutes early. He sat in his usual seat, next to where Henry sat, nervously awaiting him. Ten minutes went by. Riffy stared at the door. Finally, the door opened and Henry came in.

"Riffy!" Henry said, his surprise apparent.

Riffy nodded and looked away as Henry sat down next to him, taking out his notebook. A few seconds later, Riffy turned toward him and pasted a concerned expression on his face, and said, "Hold still, Henry. What's this?" He peered into Henry's ear. "It looks like there's something...." He poked a finger in, and Henry pulled back. "Oh, it's nothing," Riffy said, satisfied that the implant was in place. "Sorry."

Riffy maintained his owl-like stare to conceal his glee. Henry looked so sad and tired - Riffy hoped the transmissions from his ear implant would cheer him up.

Henry heard Riffy say "Don't let on that you can hear me." Henry looked at Riffy who continued to stare straight ahead. Confused, Henry looked back down at his notes.

"Don't look up and don't talk - this is not Riffy speaking." Henry's eyes opened wide and he glanced at Riffy again who sat staring straight ahead.

"We're using Riffy's voice to make you feel more comfortable. We have secured your family from harm and have a plan to extract you and all the other programmers. Riffy needs your help."

The words came into Henry's head as a whisper. He looked at Riffy, his face expressionless, and Riffy turned to Henry with an all but imperceptible nod. Henry had expected to die, saving his family in the process, but could this save him?

Riffy's voice in his ear continued, giving Henry instructions for how to communicate with Riffy in code, using "Riff" instead of "Riffy," among other things.

"Riff, is everything on schedule?" Henry asked, and he and Riffy began a coded conversation that sounded completely normal.

***

Over the next several days, Remi, Henry, and Riffy coordinated plans. Riffy and Remi could mouth their words while hiding their lips from the omnipresent cameras, and their Collars would translate what they were saying. Henry could hear but not speak, so his responses were sighs, harrumphs, hmm, and ugh to acknowledge Remi and Riffy. It got tedious.

One of their tasks was to identify how many guards were located at the entrances of the three facilities. Remi and Riffy mouthed their responses in bed under their covers to complete their reports, but it took longer for Henry, as he had to answer questions with guttural sounds. "Are there five or more entrances?" Harrumph.

"Are there fewer than three?" Sigh.

"Are there more than five guards?" Harrumph.

"Are there fewer than ten guards?" Sigh.

Through this tedious process though, the team mapped the three sites' defenses. Hurley hacked into satellite imagery, confirming what Henry, Riffy, and Remi had reported.

***

Looking at "pretty little thing" Moloana gave no clue that she had been one of the first female Navy SEALs. When she took control of the paramilitary operation against RICKY, she met with the top officers leading the offensive wearing a tight little grey dress, high heels, and a sweet smile. As she stood before them, the Italian militia poked each other and made crude remarks, a couple of them whistled. Moloana motioned for the officer who had been most disrespectful to come up front. She stepped towards him, placed her hand gently on his cheek, and smiled sweetly. He smiled broadly back and looked out at his friends, winking and nodding. In an instant he was on the floor with Moloana's foot on his neck. The room roared, but she didn't lift her foot. The man turned red and flailed his arms, but she didn't let up. As he struggled to breathe, the applause in the room turned to gasps. Finally, he mouthed, "Sorry," and Moloana let him go. No one questioned her authority after that.

After presenting the latest intelligence, she ordered another meeting, when final plans would be put in place. She left the room to chants of "La Cagna, La Cagna!" It's not often that 'La Cagna' (bitch) is used as a sign of respect, but Moloana relished it.

***

The RICKY bosses had made $40 million from Riffy's AI bank loan officer, but they expected to make $400 billion from their AI programmer. Rolling in cash, the RICKY bosses concentrated on the AI programmer project, stashing the profits from the bank swindles into safe places. Unfortunately for them, their implants allowed Chops and Hurley to hear everything they were doing and where most of their money was located.

Edie and Lucy stole money the from RICKY bosses, leaving money trails leading to other RICKY bosses, precipitating distrust within the organization. They listened as the bosses argued, sometimes coming to blows – the seeds of distrust were sowed.

Chapter 36 – Back on Maui

Moto loved the freedom of swimming at Little Beach where bathing suits were optional and his confidence soared, swimming without the calming pattern. Floating in the waves, he considered Otis' and Daisy Sue's energy bubble, picturing a sphere that would burst, but then come back together repeatedly, and a quantum computer floating inside doing the same thing—bursting and coming back to together. *Eureka!* He thought then he laughed, remembering his shower with Lily.

Moto was excited and full of energy, but he couldn't concentrate, so he pictured the pattern to initiate his mania, and the universe exploded. His mind calculated, analyzed, and expanded. He saw what his H_2O system needed to be and the path to get there. Computer code filled his head.

He pictured the calming pattern. "Wow!" he said out loud and his mind calmed down as he ran out of the water, got dressed, and dashed to the parking lot, calling Otis for a ride up to Kula. Fist pump!

Moto flagged Otis down from one of the food trucks near Big Beach. "You want a fish taco, Otis?" he asked, running to the car. "Here, take this one." Moto thrust the taco at Otis' face. Never one to turn down food, Otis grabbed it and part of Moto's hand, pulling him towards the car, and ate it in two bites. Moto ran around to the passenger seat, and they sped back toward Kula.

Moto talked non-stop, and what he was saying was so astounding that Otis had to stop the car and just listen to him. He talked Otis' ear off, explaining how the H_2O environment could work. Otis offered a couple suggestions that Moto accepted with a pat on the back or a pull of Otis' ear.

When they got to Kula, Moto was spent, his energy drained with his words. Otis, on the other hand, couldn't even sit down as he told Daisy Sue about Moto's vision of the H_2O system. Daisy Sue listened with interest, but soon turned to questions about how she could use chemical reactions to make Moto's vision work and they were soon deep in discussion.

***

The short message Lily received from Moto that he would be staying in Kula made her feel sad but relieved. She wanted to see him, but she was busy. Although Tilly had another job on the mainland, Lily convinced Griffin help her set up the doomsday site.

Lily and Griffin arranged for contractors and site prep locally and worked in tandem to get preferential treatment: If the point of contact was male, Lily used her sleek sexiness to charm him, and if the contact was female, Griffin took charge and sweet-talked his way into getting what they needed.

Construction at the doomsday site had to wait because much of what they needed had to be shipped to Maui on a barge— actually, two full barges filled with goods and equipment. It would take three months for everything to arrive; the first barge would leave in a month, and the second a month later.

Having done all they could, Lily shuttled Griffin to the airport, hugging him tightly. *It was nice having him around – no not nice, great – oh get a grip*, Lily thought returning to the car. On

191

her way back to the hotel, she sang along with the car radio, making up Hawaiian words to the twangy native music.

***

Moto enjoyed working with his cousin and Otis, because concentrating fully on one thing was fulfilling and less nerve wracking. Others were dealing with the RICKY war and Ball and Chain and new tech. The three of them established a routine: Otis was a morning person, Daisy Sue was late-night, and Moto was twenty-four hours. In the mornings, Otis took Moto to Little Beach, where he swam and generated ideas for the H_2O environment while Otis organized the previous day's work and planned next day. When they would return to Kula in late morning, Daisy Sue started her day.

On the fourth morning, a particularly calm one, Moto swam into the surf and spotted a jellyfish, so he put on his mask and snorkel, watching an eel dart in and out of the reef, a goatfish swimming close to the sandy bottom, and a parrotfish chiseling at the coral. A jaw-dropping idea resolving several design issues ripped into his brain with such force that his mouth opened. His snorkel floated away and when he breathed, it was all water, and he choked and flailed.

Otis ripped off his clothes and ran into the water to help, along with several other clothing-optional beach-goers. It seemed an eternity to Otis before they reached him and dragged him ashore. Moto gasped for breath and coughed up water.

"Are you all right?" Otis shouted into Moto's face, hands on his shoulders. "Are you all right? Moto!"

Moto, coughing, gurgled, "Eureka!"

Chapter 37 – The Extraction

Edie and Lucy continued their intricate plan for taking RICKY's money, moving it from one RICKY boss to another, leaving a subtle trail behind, before moving the money into Hurley's offshore account. In just three days they had taken most of the $40 million back from the mobsters - the rest was cash that the bosses had squirreled away. Once they figured out that their accounts were being drained, they demanded that their accountants tell them where the money went and when the answers came back, their distrust for one another exploded. One by one, the bosses checked on their cash, revealing the locations of their stashes to Edie.

RICKY was in disarray so Edie determined it was time for action. Moloana readied her militia. Riffy and Remi confirmed they were ready and Edie pulled the trigger on the extraction. Five minutes was on the clock.

Henry sat in the back seat as Max drove him to his site. He noticed that Max was particularly quiet this morning and as he looked in the rear view mirror at Max's face he saw him sniffling. "Are you feeling all right, Max?" Henry asked. Max shook his head and Henry saw a tear run down his cheek.

"Max, pull over – tell me what's wrong!"

Max told him that his sister Mitsy had died the night before. Henry climbed into the front seat, his hand on Max's shoulder as they talked in between the silence.

The extraction team was not aware that Henry had been delayed. The person monitoring Henry's implant had left briefly and forgotten to switch the monitoring to someone else. When the five minute countdown began, Henry and Max were still twenty minutes from the site.

They sounded an alert. The five-minute countdown could not be cancelled because the troops had already been deployed, but Edie asked Lucy how long they could delay it.

"Four minutes."

"Henry! You have nine minutes to get to the site! Move!" Edie shouted into the Worm in Henry's ear. Henry heard the eerie whisper of urgency and sat straight up.

"Max, you have to get me there in 9 minutes," Henry said, grabbing Max's thigh with his hand. Max nodded and floored it. In between yelps of near misses as they raced over curbs and careened down narrow streets, Henry came clean to Max as he bore down on the accelerator, dodging cars that seemed to lurch out of nowhere and banging over speed bumps.

When they arrived at the site, they were a minute late, and Moloana's militia was already in a firefight with RICKY. Max raced the car to a door that Moloana's militia had secured.

"You're coming with us, Max! Riffy and I've got your back!" Henry yelled as he leaped out of the car. Max followed, still in shock, and they ran inside.

Everyone was on the floor under tables and desks. Henry's first task was to tell them to take cover - check. He ran to one of the desks, fumbled with a key for what felt like an eternity and grabbed a small device. He pressed a button that sent a signal to every laptop in the room, wiping them clean and destroying Ricky's code.

They were two minutes late. Henry shouted to the room, "We're evacuating the building! You and your families will be safe. Everyone go to the south door!"

There was a slight pause while the terrified programmers cowered on the floor stared at Henry, afraid to move.

"NOW!" he shrieked. They all rushed to the door, pushing and shoving. "You, too, Max. Get going!"

Henry stood back herding everyone like a mother hen with her chicks and pleading for order so no one would get hurt. When he received the word that RICKY's re-enforcements would arrive before they could load everyone onto the helicopters, Henry told Max to make sure everyone makes it while he stayed behind to delay RICKY from stopping the escape.

Max grabbed Henry by the shoulders. "Henry, I have nothing to live for. Mitsy is gone, and I have the same disease she had. Go! Go be with your family. Let me do this for you."

Henry had to think quickly. He stared at Max, his eyes turning red, but somehow he understood Max's grief for his sister – Max didn't want to die, but he didn't care if he died, either. Henry thought of his family and his voice cracked as he whispered, "Thank you, Max," and Max ran to the front door while Henry turned to the programmers, still pushing out the back.

Henry herded the programmers through the woods to a clearing where five helicopters whirred loudly. Last to board, Henry turned around when he heard gunshots and Max's forlorn yelp. The last helicopter took off and Henry sat motionless, strapped in his seat, staring straight ahead at an image of Max's sad face.

195

When Remi received the five-minute countdown, she hurried to her workplace and looked around.

"Where's the jerk who sits there?" she called to the guard, pointing to the empty chair. "I need to ask him a question right now." The guard shrugged his shoulders. "Let me go look for him. I have a deadline!" The guard just stared at her, and she pushed through the door and ran to his room. Barging in, she saw towels hanging over the cameras and then muffled a shriek when she saw a motionless body hanging from a makeshift noose. *Just one more day, man*, she thought. *You should have waited just one more day.*

She dashed back to the workroom, rushing past the guard.

"Is he coming?" the guard asked, not even looking up.

"He said he needed a couple more minutes," and Remi hurried back to her desk, seeing the countdown at fifteen seconds. She opened her laptop and fumbled for the on switch.

She heard gunfire outside. "Get down!" she yelled. "Everybody get down!"

As the programmers scrambled to get under their desks, her Collar transmitted the virus to wipe out all the laptops in the room. Remembering that the hanged man's laptop hadn't been turned on yet, she crawled to the empty desk and opened it.

"Everybody, listen up!" she shouted. "This is a rescue. Your families are already safe. You will be safe. They'll come to get us soon."

It was sooner than soon, as the door slammed open and three of Moloana's militia knocked out the guard and ordered everyone out. The bewildered programmers rushed to the door, and the militia led them past several dead RICKY guards to the waiting helicopters. Remi took her seat in the last helicopter, taking deep breaths and exhaling in relief, her head back. *Nice job*, she heard on her Collar from Riffy.

***

Riffy's site extraction went like clockwork. He sat in his helicopter listening to Henry and Remi as they struggled to clear theirs. Riffy felt sad about Max, who had vouched for him and to whom he had promised he would 'have his back'.

One of the militiamen scanned each passenger with a wand, searching for tracking devices. When he scanned Riffy, the wand beeped. "Sir, please stand up," the man ordered. He searched Riffy more closely until he zeroed in on Riffy's backside. "Drop 'em, sir, and bend over," The man ordered. Riffy complied and after a particularly personal and uncomfortable search, the man quipped, "How'd you get that up there?"

"Pretty sure I know," Riffy replied, pulling up his pants. The man placed the device into a small capsule, attaching it to a drone, and dispatched it. Riffy watched out the window as it flew off and thought, *I wonder what Henry has up his ass.*

The drone made its way, with the help of several operatives in route, to a small, unmanned boat in the Mediterranean. RICKY was expected to follow the tracking device to the decoy boat, and if they destroyed the boat, the search for Riffy would end.

If they boarded the boat, the drone would explode. Either way, Riffy would be safe. Henry's tracking device, implanted in the back of his neck, took a leisurely ride down the Danube.

The fifteen helicopters flew to fifteen different destinations, all within twenty miles of the extraction sites. Moloana had determined that the shorter air time and multiple destinations would make it more difficult for RICKY to track them.

The escape plan was the biggest expense, filled with misdirection, separate routes, and weeks of hiding in multiple locations before they would reach their final destinations. Eight driverless cars in a caravan headed toward Spain, filled with food wrappers and empty drink cups, along with fingerprints and DNA from the evacuees. Once RICKY detected the caravan, Hurley had the cars split up to out-of-the-way locations where they were parked for RICKY to find, triggering a manhunt that wouldn't find anyone.

The evacuees left their first stopover a few at a time, accompanied by one of Moloana's militiamen and were transported to secondary sites in Amsterdam, Belgium, Innsbruck, Munich, Zurich, and the doomsday site outside Paris - six to eight people at each location.

They would stay for up to three weeks before heading to their final destination, but not before they were debriefed and allowed to speak with their families. Sophie Jean and Chops interviewed each one to determine to what extent they could be trusted, while Hurley did additional background checks on the evacuees and their families.

***

Six bosses had been killed over the money squabble, but a few had formed alliances. Edie and Lucy continued pitting bosses against one another to minimize the threat against the evacuees *and* Riffy.

In Munich, the cars drove up to Brandy's restaurant, and he met them out front, standing and smiling with his imposing form. He walked smoothly to the car, his pain invisible to everyone but him. He handed each evacuee a pair of glasses, which they put on before stepping to the curb.

Each pair had cost $40 thousand to manufacture. The technology included an infrared device that blocked facial recognition – critical for hiding the evacuees from RICKY. The lenses were thin and light but assembled like a cream-filled cookie - inside the glass lens was a jell-like substance programmed to simulate bifocals or trifocals. The lenses corrected color blindness, and were pre-adjusted to fit each person's head, face, and visual needs. The glasses even improved the long-range and close-in vision of the three people who did not need to wear glasses to encourage them to wear them all the time.

Ball and Chain was developing a similar product that sensed what the iris was looking at, auto-focusing the lenses. This technology would far exceed the capability of the human eye, providing telescopic vision of 20X, turning 20/20 vision into an extraordinary 20/1. Future plans included night vision and the detection of wave lengths higher than visual light, allowing the wearer to see through some solid objects. On a parallel development path, the technology was being made available for lens implants, using simple cataract surgery. Ball and Chain already had patents on the eyeglasses and the eye implants and Moto expected this product to increase revenues by a trillion dollars within three years.

***

At the Egg Roll Inn, Brandy instructed each evacuee to enter the last stall in the wood paneled restroom, shut the ornately carved wooden door, and wait. A panel opened and the evacuee would walk 200 yards through an underground tunnel that opened into a large room with twelve sleeping alcoves. When the last evacuee reached the bunker, Brandy raised a stein and shouted, "Let's party!" It didn't take long before everyone was filled with lumpia and bier. Brandy announced that they would be having their final destination interviews the next day, which raised a babble of excitement and curiosity, but the day's events took their toll and everyone crawled into their soft, comfortable sleeping quarters.

***

Chops and Sophie Jean interviewed each of the evacuees in Munich for about two hours using a script specifically tailored for each person. The interviewees were given three options for a final destination, but in reality the decision had already been made, so Chops and Sophie Jean subtly led each one to choose the correct option.

Brandy felt like a papa to his evacuees, so after their interviews, he spent time with each one to review the option he or she chose, confirming and supporting their choices. Then, over the next three days, one by one, he bid them auf wiedersehen as they left the Egg Roll Inn.

After the last evacuee left, Brandy limped into the restaurant and sat down. His pain was unbearable, even overshadowing his love and friendships. So, in the next few days, he put his affairs in order. Then, late one night, he washed down all the pain pills he had saved up with his favorite bier and shots of schnapps. He lay back in his favorite chair and smiled as his pain drifted away and he floated to the heavens.

Chapter 38 – Success on the Valley Isle

Moto jabbered incessantly while Otis drove them up the mountain, but Otis was distracted by the road, slippery from a recent rain, and still a little shaken about Moto's safety after the "water-swallowing" incident.

Otis slammed on the brakes.

Moto had gotten through. "That is fucking brilliant!" Otis yelled. Moto smiled and slapped Otis on the arm with the back of his hand, eased back into his seat, and closed his eyes while Otis drove on to Daisy Sue's in open-eyed silence.

When they arrived, Moto took a nap and Otis sat at the work table alone, waiting for Daisy Sue. *Moto's the driving force, but I'm the one that can make this happen*, Otis thought, *I understand chemical engineering and programming and quantum systems.* Otis sighed and plopped his head onto the table but his eyes snapped opened when fear of failure tried to gain a foothold, but closed again when he realized that he had the best support team in the world cheering him on.

Daisy Sue joined him. "Hi, Daisy Sue," Otis said, in all innocence as he gazed around the room, now leaning back in his chair.

Daisy Sue smiled broadly and said, "Tell!"

They talked, giggling like school children, about Moto's breakthrough and how to translate it into real terms and make it work.

Meanwhile, Moto dreamed about his quantum computing jellyfish that stayed together in flowing water. He pictured it reconstructing itself when it was broken apart by reefs or other objects. He dreamed about Daisy Sue holding the jellyfish like a handful of pizza dough and that Otis was floating alongside, pointing to how the jellyfish quantum computer broke apart and reconstructed itself. When he woke up, he joined them. "Got it figured out yet?" Moto quipped.

"I think we do," Otis said in all honesty with a grin and a wink to Daisy Sue and he gave a long dissertation of how the individual "cells" of the quantum computing jellyfish would have universal functionality. Each cell would be the size of a flea but would contain a homing signal, a chemical signature, universal quantum code, and 200 times the knowledge it needed to perform its individual function. A fully functional jellyfish would be the size of a ping pong ball, held together with a chemical bond and a powerful energy field. But if broken apart, it would reassemble at 95 percent needing only a third of its cells.

Otis finished his dissertation and Moto stood up saying calmly, "They need to be self-replicating." Daisy Sue cocked her head inquisitively, but Otis' eyes lit up. The bodies of water around the world were vast, so they needed millions of quantum jellyfish.

"Otis," Moto said, "I asked Sophie Jean to find a backup for you. You are the most important person on the planet right now because this tech is a key to our survival. We can't afford to lose you for any reason, but if we did, we need someone who can step in and do what, right now, only you can do"

"I won't let you down," Otis said with a straight face, "I'd hug you but I'm too important," and they laughed.

The three of them spent the next two days in intensive discussions until Otis announced that they were ready to start development. Moto asked Otis to take him back to the hotel, saying he'd be back in a few days with the specs.

***

Moto opened the hotel door, seeing Lily laughing with a very pretty, young, blonde woman, an almost empty bottle of champagne on the table between them. "So, you're Moto, huh?" Lily's friend said in a sweet, sultry voice.

"This is Honey, Moto," Lily laughed. Moto stretched out his hand, but Honey jumped up and hugged him playfully.

"We're celebrating tonight," Lily said, still laughing as she looked at Honey, whose arm was draped over Moto's shoulder. "The barge with all our supplies came in today, a month early, and Honey has agreed to be our doomsday site manager on Maui."

Moto turned to Honey and hugged her back, picking her up off her feet and swinging her around. "Welcome to the team, Honey!" he said in the same sultry, sweet voice she had used on him.

They all laughed and chatted for a few minutes, but Lily noticed that Moto had become quiet and withdrawn from the conversation. "So are you about ready to get to work, Moto," Lily asked calmly. He nodded and went to the corner to work on his quantum jellyfish program specs. He pictured the pattern that would make him manic, and Lily, Honey, and the rest of the world disappeared.

"Don't worry, Honey," Lily all but shouted. "We can still celebrate. Moto won't hear a word we say." Moto, his face close to his work, didn't react.

Honey and Lily continued partying and discussing the site setup, until Lily's eyes drooped from the champagne and the long day, saying, "Bedtime, Honey," and Honey reluctantly left. She was almost to her car when Lily ran up to her from behind with a Collar. "I forgot to give you this, Honey," Lily said. "We'll talk tomorrow, but put it on and never take it off." Honey put on the Collar and drove away, hearing Lily whisper in her ear, "Sleep tight!"

When Lily awoke the next morning, Moto was still at it. Now that he could control his mental states, Sophie Jean recommended he sleep at least four hours a day, so Lily tapped him on the shoulder, getting a sliver of his attention, and asked him to come back to his normal state. Sighing with impatience, Moto pictured the calming pattern and let Lily lead him into the bedroom, falling asleep muttering, "No more than four hours." Lily set the timer and went back to sleep herself.

When the timer went off, Moto slipped out of bed and went back to his laptop, returning to his manic state to continue his work. Lily ordered fish tacos and after eating more than half of them, hand-fed Moto the left-overs.

She left quietly to meet with Honey at the doomsday site and calculated that Moto would run out of energy in nine hours. She looked forward to coming back to feed her baby bird calamari, mahi mahi, or maybe coconut shrimp...yum!

Lily arrived at the site and Honey was outside doing cartwheels, laughing and giggling. Not to be outdone, Lily got out of the car and did six cartwheels ending up directly in front of her.

They laughed walking arm in arm to the entrance. There was a humming sound coming from within. Honey gushed, "The thermal generator is on line, Lily! Haleakala will provide us with enough power for centuries! All the deliveries from the barge have been unloaded and organized! We're all set for the rest of the construction work!" Honey beamed.

Walking through the site and seeing the progress, Lily wasn't sure whether she was more impressed with Honey or with herself for hiring her. "Honey Bee," she said, her hands on her hips, "I'm gonna miss you when I leave tomorrow. But you don't need me here anymore," looking around proudly. "Yeah, I'm gonna really miss you." Honey smiled broadly and did a cartwheel. Then Lily did one, two, and then a third. "Two in Alabama, one outside Paris, one in Costa Rica, and one in Zimbabwe," Lily said, pointing east.

"Huh?" Honey managed, mid-cartwheel.

"That's where the other doomsday sites are. I think the weekly call is at 7:00 in the morning Central time, but you'll get a wakeup at about 1:30 AM Hawaii time in a couple of days. Maybe you can convince them to move it." Then she added, teasing, "You'll like Buddy - he's a doll."

"Thanks, Lily. You're a real sweetheart," Honey said with a wink. "I'm going to miss you, too."

"You're the bee's knees."

They both smiled, and Lily turned and walked away with a small wave of her hand, while Honey returned to assess what needed to be done and thought, *I got this*.

***

Lily, eyeing a room service cart filled with food she had ordered earlier thought, *they must have just left this*. After digging in herself, she rolled the cart over to Moto, feeding him enough to last another six hours. Lying on the couch, just as she drifted off to sleep, she heard Moto mumble, "Thanks, Lily."

At midnight, Lily woke up, and made Moto go to sleep for another four hours. And the cycle continued.

Chapter 39 – Where's Shorty?

Buddy was excited about the prospect of welcoming permanent residents to the Paris doomsday site, but he had to play it cool because he was, well, cool. He hung back as his new tenants arrived, greeted by his second in command. Teddy was a young, energetic man with light brown curly hair, offering each new arrival a warm handshake and a smile to help the guests, whose lives had been upended, feel comfortable. Most of the six evacuees were older and were happy to live out the rest of their lives in secure comfort. Three had spouses who had arrived the day before, two were single, and one, Maggie, was a strikingly beautiful young French woman of about Buddy's age who had white curly hair. The report on Maggie was inconclusive about her trustworthiness: She might be trustworthy but had to be tested and so Buddy put her in a comfortable setting while monitoring her.

Buddy couldn't take his eyes off her, but Maggie couldn't take her eyes off Teddy. As Teddy walked over to greet Maggie, he reached into his pocket with his left hand and slipped on a wedding ring and then scratched his forehead so Maggie could see it. This was a test of her trustworthiness that Sophie Jean had devised and Buddy was encouraged when Maggie turned off the charm and formally shook hands with Teddy.

Buddy was also encouraged in a different way, considered introducing himself, but his need to be cool and do his job overrode his desire to sweep Maggie off her feet, so he stayed back. Buddy was a company man above all, meticulous about rooting out even the smallest risk to his doomsday site. Sometimes he developed risk-mitigation plans for things that could never happen, just because he enjoyed it. He hoped Maggie would pass the test but wouldn't let his desires compromise the mission.

The Paris site was first to house permanent residents so Buddy became the "go to" guy for the other site managers. He was driven to be respected and valuable – ambition and power grabbing were not in his nature, though. Some people who want respect react to failure or criticism with anger or frustration, but Buddy's reaction was to accept responsibility and work harder. Sophie Jean considered Buddy a remarkable person.

***

Lucy rushed to pack her bags to get to Amsterdam. Shorty had gone silent. The evacuees in Amsterdam were at the safe house, but Shorty was missing. Lucy got up to speed on the six evacuees during her three-hour flight to the Netherlands and when she arrived, she gave her cab driver the address of one of the brothels in Amsterdam's red-light district. Pulling up to the front door of the brothel, the cabbie smirked and said, "New job?"

Lucy stepped out of the cab and paid the cabbie for the fare through the window, saying angrily "Yeah, stop by sometime and I'll give you your fucking tip."

Lucy walked around to the back door and went in, using her key. She had instructions to go to room 4 and enter the safe room through an armoire. Room 4 was empty when she entered and she dashed to the armoire and inserted her key card in a slot hidden near the bottom. A crawlway opened and she muttered a curse as she got down on all fours. Once through the opening, Lucy stood up and followed a set of stairs down to a large room with doors to private spaces for the evacuees. When she entered, they looked up in alarm.

"Hi, everyone, I'm Lucy. Shorty had a personal emergency, so I've come to help you get to your final destination." Sighs of relief filled the room, and the evacuees smiled, sitting back in their chairs.

"Shorty left early this morning," one of the evacuees said. "He looked confused and scared and though he walked right by me, he left without saying a word."

"Yes, he asked that we come and help out because his grandmother passed away," Lucy lied. "I have all of your information and we will be going forward with your final reviews during the next two days." She looked around the room at the relieved faces and added with every ounce of conviction she could muster, "I'm so happy to be here!"

***

It was hard for anyone in the world to hide from Hurley, and he pulled out all the stops looking for Shorty but the trail went cold when Shorty walked out the door that morning. His cell phone and tracking device were silent, no hits on his credit cards or facial recognition, and no activity on his financial accounts. It was like he went "poof!" two steps out the door of the safe house.

Edie was worried. Already upset about Shorty's unapproved communication with the team, his disappearance increased the risk to Riffy and the other evacuees. Riffy had gone to Belgium, and Edie wanted to go there to ensure his safety, but Chops told her to stay in Assisi and monitor RICKY communications.

"Edie!" Riffy's voice shouted into her Collar. "What you are doing is critical, suck it up, babe," he ordered. "I'm headed back to the States tonight. See you in eight days."

"Thanks, Riff," Edie replied. "I'll clean up here and we'll have a pleasant chat when I get back."

Riffy laughed. That "pleasant chat" was a euphemism for one of two things, and he was pretty sure he wasn't in the doghouse. Still, the lie that he was going home that night might put him in one when Edie found out he was staying in Europe until the rest of the team went home.

Hurley was exasperated that he couldn't find Shorty. The lack of any communication about Shorty from the RICKY bosses should have been a comfort, but it was just the opposite. Hurley needed an explanation and there wasn't one. Something unknown was at play, so he decided to talk through the problem with Sophie Jean before he called an emergency meeting with the whole team.

It was the middle of the night when Sophie Jean's doorbell rang, waking her up. At the door Hurley stood in all his casual splendor, looking concerned. She beckoned him in, and Hurley was talking about what happened to Shorty even before he reached the couch and sat down. He sounded disjointed, so Sophie Jean held up a hand to stop him and led him through a more logical conversation. As he answered each of Sophie Jean's questions, she became increasingly unnerved. Hurley was right: something else was at play.

Sophie Jean suggested that Hurley expand the background check to twenty years from the usual ten. He calmed down when he had a direction to take and exhausted, he leaned back and closed his eyes. Sophie Jean resisted moving closer, instead finding a pillow and a blanket for him and going back to her bedroom to finish her night's sleep.

When she woke up, Hurley was lying in bed next to her, snoring loudly. Comfortably surprised, she dressed, made breakfast, and left a note on the counter telling Hurley to eat before going back to work. She left reluctantly and felt distracted all day, waiting for the call from Hurley that never came. She left early and when she walked through her front door, Hurley was still there, working. He didn't look up.

"Honey, I'm home," Sophie Jean said with a tinge of sarcasm. Hurley didn't respond. She went into the kitchen and made a pizza, and when it came out of the oven, Hurley hadn't acknowledged her yet, so she took a piece and held it under his nose.

"Is this vegan?" he asked without looking up.

Sophie Jean sighed and rolled her eyes. "Yesssss," she replied and then added firmly, with a sly grin, "but not gluten-free." Hurley ate five slices while tapping away at his laptop, mumbling "I could get used to this." Then he finally looked up. "There's no logical explanation," he said shaking his head and sighing.

Sophie Jean sat down next to him and touched his arm. "There are a finite number of logical explanations, but an infinite number of illogical ones," she said quietly. "What are the clues?"

Hurley sighed. "There's only one clue: the eyewitness at the safe house that saw him leave early that day. He said that Shorty seemed distracted and in a hurry."

"And you haven't been able to find out why he left in that state?"

Hurley shook his head. "I monitored all his communications and the communications of all his friends and family and came up with nothing. He had been 'dark' like he was supposed to, and nothing financial was out of the ordinary."

"And his tracking device?"

"It stopped reporting just thirty feet outside the safe house. I can't explain it. The tracking device technology is supposed to withstand anything except a total cataclysmic event," Hurley trailed off. After a moment, he added, "There must be a technology that we don't know about that either communicated with him or compromised his tracking device."

Sophie Jean thought for a moment. "These events are outside our paradigm. We need the tracking device developers to theorize about what could cause a complete failure, and I think we should ask Indy to contact his grandmother, Bodhi, to give us some clues."

Hurley turned from his laptop and looked deep into Sophie Jean's eyes, "What if I missed something?"

Sophie Jean gently grabbed both of his ears and whispered, "Then you'll look foolish, but who the hell cares?"

Chapter 40 – The Search for Shorty

Chops got down on his knees at the Fishy Fishy restaurant, holding a small, black-velvet-covered box, and opened it so that Daisy could see the sparkling ruby red engagement ring. She looked at him and nodded. Standing, she grinned, pulling him to his feet, embracing him as the patrons cheered. Chops placed the ring on Daisy's finger and the room roared. They walked around, giving high fives to all the patrons and for the next couple of hours they celebrated with champagne and the best that Fishy Fishy had to offer. When the bill came, Chops opened an envelope that came with it and found $800 the people in the restaurant had contributed, so he left $200 for the bill and a $600 tip. They smiled warmly as they walked hand in hand out the door.

When they reached Daisy's apartment, Chops took his Collar off "emergency only" mode and saw Hurley's messages about Shorty's disappearance. When Daisy leaned down and put her arms around him, he said, "This is the happiest and most frightening day of my life." Daisy just had time to pull back before Chops added hurriedly. "No, I'm not afraid of our future life together! I am afraid of the consequences of what happened in Amsterdam a couple of days ago."

Daisy smiled. "Even if you were afraid of our relationship, I would make sure it worked."

Chops kissed her on the nose, looked lovingly into her eyes, and said, "I have to call Sophie Jean." Daisy laughed and went into the bedroom. *I'm the luckiest man alive*, he thought.

Sophie Jean was leaning in to kiss Hurley when she got the call from Chops. "Are you engaged yet?" she asked.

"Yes, but that's not why I'm calling. What's going on with Shorty?"

"Things are moving forward. There's nothing urgent you can do right now, so go enjoy your engagement night. I'll call you in the morning."

Chops sighed and stumbled to the bedroom, his head spinning with the catastrophic scenarios that might unfold, but when he saw Daisy lying in bed, her head propped up on one arm, smiling – *poof* – his worries vanished.

***

The report from the developers analyzing what might have caused a catastrophic failure in Shorty's tracking device contained few possibilities. Beethoven reported that the satellite feeds from spy satellites he'd hacked showed Shorty walking out of the safe house, turning a corner into an alley, but no sign of him after that. Beethoven received a thumb drive from a contact that had hacked a Chinese surveillance satellite, and he watched the image as it was uploading into his computer. For a brief moment, he saw a flash of light over Shorty, but when he reviewed the uploaded video, the flash of light was gone. He figured it was just a glitch, but mentioned it to Hurley in passing.

Hurley rushed to Beethoven for the copy of the thumb drive so he could go through the same process. He had the same result: He saw the flash of light during the upload, but when he reviewed it after uploading, the flash of light was gone. He copied the thumb drive again and brought Sophie Jean in to repeat the process - with the same result.

They stared at each other in disbelief, but then Sophie Jean put her finger to her lips and motioned them to follow her. As they walked through the office, she gestured to as many of the core team she could to come with them quietly. When they arrived at the safe room and had gone through the protocol, Sophie Jean began, "You have all heard about Shorty's disappearance in Amsterdam. There is a distinct possibility that there is a technology at play that we are not aware of."

Hurley, Beethoven and Sophie Jean went through every detail of Shorty's disappearance; ending with Sophie Jean's summary that "we have a repeatable anomaly that cannot be ignored."

They agreed that everyone that could be brought in should be, but they were not sure if the electrical grid network was compromised, so Sophie Jean tasked Hurley with developing a test environment to see if it was. Ginger reminded Sophie Jean that they had a backup plan for communicating with Moto and Otis - the post office. Sophie Jean nodded and said she would send a hand-written message to Moto, Lily, and Otis via snail mail.

Hurley looked around the room at the "deer in headlight" expressions thinking that although prepared for the eventuality of an AI takeover, coming face to face with it was still a shock. He ordered, "Game face on" and left.

***

Daisy Sue handed Moto the letter from Sophie Jean that had come in the mail. Moto felt a sense of dread as he read it out loud. Looking up, Moto announced, "We have to go back to the mainland." Otis and Daisy Sue looked at each other and then back at Moto. "Daisy Sue, I know you don't want to leave Maui, but we need you and Otis to continue your work together, and I need Otis with me."

Daisy Sue nodded, thinking she could finally justify Moto's years of funding her research. Moto Collared Lily and asked her to make flight reservations and three minutes later, she replied that their flight would leave in 97 minutes. After ninety minutes of frenzy, they boarded the plane with a leisurely 7 minutes to spare.

Lily snuggled up to Moto for her blissful 5½-hour flight to the mainland, while Moto spent the time reviewing Hurley's reports, Lily's head on his shoulder. Daisy Sue did the same as Moto, but Otis' head leaned towards her shoulder, snoring softly.

Branco picked them up at the airport, and they rushed to the safe room, where Hurley, Sophie Jean, Beethoven, the Gingers, and Indy were deep in discussion. Moto managed a smile and with a nod to the team, introduced Daisy Sue as his cousin, a chemical engineer and key participant in their newest tech. They took a seat along the side of the table.

Hurley stood at the head and presented a summary before he started through the first report saying, "When we tested the sidewalk where Shorty went off the map, we found—"

"Let me guess," Daisy Sue broke in. "There was an area with a radius of two to four feet where the concrete showed slightly elevated but still low amounts of vanadium, calcium and magnesium."

Hurley looked surprised but nodded, adding, "There was also trace amounts of manganese and neodymium.

The shocked look on Daisy Sue's face caught Hurley's eye, but he went on to mention that there was a pile of dog poop somebody didn't clean up.

As the room began to snicker, Daisy Sue jumped to her feet. "Get me that shit!" she said sternly. Hurley looked at Moto.

"You heard the lady," Moto said.

Hurley ran out of the safe room and called his team in Amsterdam, who had picked up the poop, but thrown it away. "All of it?" Hurley asked, panicked.

"Well, there is still some residue left on the concrete."

"Get it. Cut out the concrete with the crap on it and have someone hand-carry it to me," Hurley directed.

"Yes, sir."

"Hurry! Who knows when it might rain?" Hurley rushed back to the safe room, breathless.

"The shit's on its way!" he reported and Otis snickered.

Daisy Sue said, "No, really, I mean it. We need this shit," and the room, including Daisy Sue, slowly exploded into laughter.

When it quieted down, Daisy Sue, trying to be serious, said, "The poop will tell the story."

"No shit!" Otis quipped, bursting the room into laughter again.

Hurley dismissed the meeting, but Moto hung back with his cousin. "Thanks, cuz," he whispered.

"You're welcome," she said, tugging his ear before getting up, leaving Moto alone in the room. He put his head down on the table, isolated from sound, technology, and people, and pictured his calming pattern, falling deeply asleep.

The next morning, Lily asked if anyone had seen Moto. Indy was quiet for a moment and then said, "I'll go get him. I know where he is. He's fine."

Indy dashed back to the safe room and sat at a chair across from Moto, who was still sleeping. He closed his eyes and asked his grandmother, Bodhi, if it was time for her to contact him. He pictured her face and heard her whisper "Not yet." Then he closed his eyes and reached into the dream world where he saw Moto swimming happily in the ocean, mimicking jellyfish. Indy gently called him to shore.

Moto raised his head and smiled at Indy. "Thanks, man," he said, stretching. "One day you need to show Lily how to do that," and they walked out together, Indy's arm around Moto's shoulder.

***

Moto had provided the specs for the jellyfish design before they left Maui, so Otis pulled in Indy, Weetzie and six others to begin development. He was empowered to move forward quickly and had the authorization to bring in anyone he felt he needed.

Otis reached for a chair, sitting heavily, his forehead glistening with sweat. He put a hand to his chest and took deep breaths, looking around the work area anxiously. Just then, Sophie Jean appeared standing next to him, placing her hand gently on his shoulder.

"Otis, how, ya' doin'?" she asked softly. He looked up with his big brown eyes and shook his head. "You know," Sophie Jean continued, "sometimes we get in our own way when we're trying to get something really important done, and what you are doing is critical to the future of humanity. But we're all in the same position and it takes our entire team to be successful. You know what has to be done - just relax and do it." Otis' breathing became less labored. "And stop drinking coffee for a while," she added, patting the top of his head. "You won't need it during this project. Your adrenal glands will take over."

"Okay, decaf then," Otis said, looking down.

"But if you find yourself getting anxious—after all, there's a lot riding on you—just come by for a little break. I'm always here … unless I get depressed. Then you're on your own."

Otis looked up, startled. Sophie Jean had just told a little joke, and he had never heard her tell a joke before. He broke out into a broad smile, but her face remained serious. "I mean it, Otis," she said a sternly. Chastened, Otis stopped smiling, but then Sophie Jean smiled. She had just told another little joke, but this time they laughed together.

***

When Daisy Sue's shit arrived, she led the delivery team to a conference room that she had set up as a make-shift laboratory.

220

As she opened the package and her olfactory senses reacted, she thought about how she missed her pets back on Maui. But once the aroma filled the room, the delivery team quickly left her alone to perform her analysis.

After some specialized tests, Daisy Sue sat back in her chair, shaking her head. *I've never seen anything like this*, she thought and brought in the one person who could understand – Otis. Otis stood stiffly as Daisy Sue reviewed the test results in excruciating detail.

"Very interesting, but can we take this to a different room," Otis said, waving a hand in front of his nose. "It really smells bad in here."

Daisy Sue looked at Otis with exasperation, her hands on her hips. "Otis," she said. "There are things I need to show you!" She rummaged through her purse and came up with nose plugs. "Here. I use these for swimming," she snapped, handing them to him.

Seeing her frustration and urgency, Otis thanked her and put the nose plugs on. *Much better*, he thought.

"Comfy now?" Daisy Sue asked and Otis grinned. "Good. Now look here." She brought up a chart that showed several molecules and said, "There are molecules in this sample that don't exist in nature. I ran the tests seven times with the same results."

Otis gave her a quizzical look and then blurted out, "What the fuck is *this*?"

"Exactly," and they stood in silence.

"What kind of energy would be needed to make this?" Otis finally asked.

Daisy Sue looked up. "You're right," she said, hurrying back to her lab set-up. "These molecules can't have been created from a chemical reaction alone. It would have required a significant energy force to form them." She peered through a microscope and waved her hand dismissively, and said, "You can go now."

"You're welcome?" Otis said, knowing she wouldn't hear him.

Daisy Sue continued her analysis with a new perspective, trying to determine what type of power could create those molecules. At one point during the next two days she spent working on the puzzle, a thought flashed through her mind about Otis and how grateful she was for his insight, but it was more important to continue - she'd thank him later.

On the other side of the world, the RICKY evacuation was on hold. Except for the doomsday site near Paris, which was both a safe house and a final destination, the evacuees had been told they might have to stay put for up to two months to ensure their safety, but the stockpile of supplies would last for only four weeks, so it was urgent for Daisy Sue to complete her analysis and for the team to resolve Shorty's disappearance.

Edie continued monitoring the RICKY bosses' communications, doing a fist pump each time she heard an argument or disagreement over money between the crooks. So far, she had grabbed $80 million from the RICKY bosses, twice as much as RICKY had made from the Andy Inch bank manager scheme. Edie hoped to reimburse everyone BOD had ripped off. Although there was over $200 million allocated for the evacuation, Edie was only going to spend $38 million. *I made a $42 million profit!* She thought. *Whoo-hoo!*

RICKY's bosses' bosses were just finding out about the fiasco, and Edie listened in as they squirmed, pointing fingers at each other. The Russian mob and the Nazis were still talking as if they trusted each other, and Edie became concerned when the Nazis posited that it might have been an infiltration that had disrupted their plans, rather than the other members of the alliance. They mentioned Henry, Max, and Riffy as possibilities, but since Henry had been coerced into the position and Max had been killed, they named Riffy their prime suspect.

Edie broke out in a cold sweat when the Nazi suggested that maybe Riffy didn't die in that boat accident. "Nyet, we have confirmation," The Russian mob boss replied.

"Well, let's see what we can find out about their background," the Nazi said. "They couldn't have been working alone."

Edie listened as the Nazi and Russian mob bosses broke away from the others and continued a "private" conversation where they discussed how, even though all of their data and backups had been destroyed, they had preserved and hidden a hard drive with all the programming on it. Unfortunately for them, Remi and Riffy had modified the program code to look like it worked, but in reality, it would compromise not only the people who were trying to use it but also anyone they communicated with. Edie couldn't contain a smile as the bosses congratulated each other for having "foiled" the plan to destroy all the work that had been done. She listened as they spread the news up their own organizations, enjoying congratulations, while keeping it a secret from the rest of RICKY.

Edie decided to cause the RICKY bosses even more problems, so she started a rumor that one of the Nazis had the AI code and was willing to provide it for a fee. One by one, the RICKY bosses contacted the Nazi boss, first threatening him and then bribing him for the code. The Nazi was an easy target; Sophie's Jean's analysis was on point.

Edie sat back and let the drama unfold as the in-fighting continued. *Just wait until they see what they've been fighting over*, she thought, giggling out loud.

Chapter 42 - Shorty's Murderer Exposed

"Otis!" Daisy Sue yelled, so loudly that the many in the office looked up.

Otis ran into Daisy Sue's lab but before he could catch his breath, she shouted "I know what happened, but no one in the world has this technology."

Once Otis closed his gaping mouth, he sat down and said "OK, let's start from the beginning."

She began with Chemistry 101 and ended with hydrogen fusion, stepping through exactly what would have had to happen to leave the chemical residue on the concrete. For every question Otis had, Daisy Sue showed him three test results as an answer.

Finally, Otis held up his hand, "Daisy Sue," he said, calmly. "I understand. We need to bring in the rest of the team."

She nodded, and Otis called an emergency meeting in the safe room. They walked into the meeting together and Otis leaned over whispering, "Great work." Daisy Sue walked straight ahead with a determined look on her face and confidence in her heart.

Everyone arrived quickly and once settled, Otis explained that the room's security had been modified eliminating electricity; skylights that amplified sunlight, but blocked EMF signals, provided the lighting. Otis got to the point and said, "Daisy Sue has finished her analysis. We need you to help us figure this out."

He sat down and deferred to Daisy Sue, who walked through the report, without going too deep into the technical chemical analysis. When necessary, Otis chimed in to provide a layman's explanation of what she was saying.

"The bottom line is that something caused a mini hydrogen fusion reaction that cascaded to all the cells in Shorty's body. The chemistry of what *was* left behind and what was *not* supports only that. There is no technology that I'm aware of that can cause this. We need your help," she reported.

The room fell silent. Moto's face turned white and he asked everyone to give him a few minutes as he pictured his manic pattern. He exploded into a non-stop analysis of the last ten years of Ball and Chain's technical development, but the team couldn't follow what he was saying. Lily touched his arm and asked him to picture his calming pattern.

He took a deep breath and closed his eyes. "In the early days of Ball and Chain," he said, "we were developing a method for creating cold fusion. We weren't successful because our client stopped the funding, but I felt that we could have eventually found a way to use chemicals and concentrated waves to fuse two hydrogen atoms and produce power. Other methods that were being tried around the world forced two atoms together with great power, but we thought it could be done with only minimal power. That research is locked in our secure database. Is it possible that someone hacked our technology?"

"If it helps," Daisy Sue said, "it appears that the chain reaction began inside Shorty's body and spread outward, probably in less than a second, but there must have been some energy boundary that controlled the reaction and stopped it from affecting anything else. It also appears that Shorty was lifted off the ground slightly before the reaction occurred."

Everyone except Moto looked around the room, hoping for an answer and Moto took deep breath. "What you say happened here is consistent with the technology we were developing. It's a little different, but it's too much of a coincidence to ignore that what we did could have been developed into something like this."

Hurley jumped to his feet and directed the team to see whether the projects database had been compromised. But when Moto asked them to check whether any of the newer technologies, including the electrical grid network, could have accessed the database, the bustle of activity stopped.

"But that would mean...." Hurley's voice trailed off.

"Yes," Moto replied, his voice flat and emotionless. "It would mean we have either a traitor or an infiltration."

Sophie Jean looked around the room, "You can trust everyone in this room and beyond," she said confidently, "Everyone!"

Moto looked down and nodded. Hurley leaned on the table, looking impatiently at Moto. "OK, Hurley, you know what to do," Moto said and Hurley dashed out the door.

It was early afternoon when they reached consensus on how to proceed. When Moto got home, Hurley was waiting at the front door and Moto put his arm around his shoulder as they walked in and sat down in the living room.

"As you know," Hurley said, leaning forward, "the backups for all of our technology are located in seven sites around the world. None of the sites contain all of the backups, and each project is backed up at three of the sites. The data drives are powered down except when we power them up for the annual hardware check."

Moto spun the index finger of his right hand in quick circles, motioning Hurley to get on with it.

"Well, prior to the electrical grid network, the drives were considered offline if they were powered down. But once we implemented it, we installed a series of mechanical devices that would completely disconnect the electrical grid from the drives."

Moto cocked his head to one side signaling his continued desire to go faster.

"Well, every six months, we perform a media check. We power up all the drives in all seven sites at the same time for twenty-two seconds while we check the media for hardware errors. The backups aren't accessed electronically or physically except for that test."

Moto nodded. "Did you—"

"Don't worry. We're checking the electrical grid to see if our network was compromised. It'll take several days to ensure 99.9 percent coverage of the grid. Beethoven is checking for power fluctuations and we're scanning every power cord in the world."

Moto sat back and crossed his legs. "How could—"

Hurley stopped him again. "If we don't have an answer we will scrape shit off the sidewalk to get you one." Moto was almost too tired to smile, but Hurley had such a serious look on his face that Moto broke out in laughter. Soon they were laughing like kids in a library trying to stay quiet.

"Thanks, man," Moto said looking down at the floor and then as if on cue, they leaned back in their chairs.

Chapter 43 – Success and Challenge

Chops arranged for the team in Europe to return to the west coast while the rest of the evacuees went to their final destinations. Trust funds would support the evacuees and their families in beautiful, remote locations, with enough work and activity to keep them busy and happy for the rest of their lives.

It was certain that the RICKY bosses were not involved in Shorty's disappearance, so monitoring and disrupting them became a "steady state" effort that Abbie, who took particular pleasure in toying with them, performed by secretly leaking their activities to law enforcement worldwide.

The returning team members took separate flights from Europe for security reasons but met for their connecting flight at the concierge lounge in Newark. Because they were wearing their facial recognition-blocking glasses, they squinted, unsure of each other at first. After hugs all around, Chops walked in with a tall lady approximately his own age, which they all knew must be Daisy.

Questions flew at Daisy from every direction, interrupted only when Riffy strode in and sprinted towards Edie. She shrieked, her voice bouncing off the ceiling and into every crevice of the room. Speechless for one of the few times in her life, Edie grabbed Riffy tightly, then pulled back to look into his eyes before burying her smile in his shoulder.

A voice over the loudspeaker announced that their flight was boarding, so Riffy chugged a glass of wine and raced to join everyone walking to the gate, where they turned to greet Henry who came running up from behind.

They had the first-class section to themselves. "Enjoy the flight!" Chops called out, smiling his funny, toothy smile. "Tomorrow morning will be back to work big time. You've all done heroic, important work." Edie winked at him and then put her arms around Riffy, begging him to dance. They stood in the aisle and rocked slowly to music only they could hear.

"We have to turn off the music now," the flight attendant whispered and Edie and Riffy laughed, sat down, and buckled up. Chops and Daisy took over as flight servers, leaving the real flight attendants to work the rest of the cabin. The celebration continued for four hours as they headed toward the west coast and home. The team was happy for the chance to relax, drink a little, and revel in the pride of their accomplishment.

When they emerged from the gate, Moto and the rest of the team were waiting with big smiles to welcome them home. After a raucous reunion ending with everyone pushing their arms up as if to raise the ceiling, Moto said in a throaty whisper that pierced the air and could not be ignored, "Thank you! Thank you all!" They left the airport, arm in arm, closer than ever. In the back of their minds was tomorrow's challenge, but the warmth they felt as a family would sustain them through the night.

***

The next morning, Sophie Jean smuggled breakfast into the safe room and the conversations were warm and pleasant, but full of anticipation. The meeting summary brought everyone up to date to the stark reality that the next challenge required a new focus.

Hurley nodded to Beethoven, who flashed an image on the team's Collars. "We found this molecule," he began, "and we're pretty sure the electrical grid network has been compromised, though we're not sure how or who compromised it. No one in the world is remotely close to developing this type of technology."

"Yes," Ginger said. "Yes," Ginger echoed in support of her sister, "Nothing like it."

Hurley stood up, "Here is what we know: A technology has been developed that can vaporize a human with an internal hydrogen fusion chain reaction that is so sophisticated that it doesn't leave a trace of evidence."

"Do we know how the electrical grid was compromised?" a quiet voice asked.

"Well, we can conjecture," Hurley replied, "but, no, we still have to follow the evidence."

Everyone spoke at once with an idea or a question, so Sophie Jean stood up and said, "Focus, people! Hurley's right—we have to follow the evidence. We can't just guess."

Chops raised his hand and called above the din, "How can we help, Hurley?" The room quieted. Chops still had it—he still had command.

"Here is what we need to do," and Hurley spent the next few minutes explaining the path forward. When he finished, Sophie Jean cautioned the team to continue their normal communications using their Collars but not to use them for any discussion about the path forward, as there was a risk that their communications were being monitored.

"If need be, we can turn our Collar communications into a weapon," Moto added.

They split into four teams. The first team was responsible for determining whether any of the backups had been accessed. Initially, they would concentrate on computer code that was associated with the initial fusion development. They would also target the code used to launch virtual quantum computers into the electrical grid. The next hardware backup check was scheduled in just five weeks, so they had to work quickly if they were going to insert code in the backups that could drop some breadcrumbs.

Team two was responsible for analyzing information from the electrical grid. Beethoven, new to the inner circle, was made team lead and, when announced, given a standing ovation. He tried to suppress a smile, but couldn't. Beethoven's life had been a struggle with self-confidence, and this assignment went a long way to correct that.

Otis and Daisy Sue headed up the third team, focusing on the technology that vaporized Shorty, and included Moloana who could think outside the box. This team had authority to pull in anyone from the other teams that might have expertise they needed.

The final team of Moto, Hurley, Ginger, Ginger, Branco, Indy, Chops, and Sophie Jean would pull all the information together, provide a strategy and management. Initially they would use a hands-off approach and get more involved as the analyses progressed—but for now Moto looked in on the team that was developing the jellyfish quantum computer and prompt them to work faster, because if the electrical grid network was compromised and all their backups had been accessed, they needed a new technology for communicating and computing.

Chapter 44 – The Slumber Party

Beethoven was being run ragged and yearned for the days when he was an unknown programmer with potential, but his expertise and knowledge of the environment had become critical. Sophie Jean realized that he was an important resource without a backup, so she reviewed the current staff and noticed an up-and-coming teen-ager with great potential.

She called Beethoven into her office, and he quickly ran in and hopped on a chair. "You're an important part of our team now, Beethoven," Sophie Jean said, leaning over her desk and smiling.

Beethoven beamed.

"Because you're so important, you've also become a risk because, without you, we could fail." She paused to allow this to sink in.

"I have a suggestion," Beethoven said. "There's a promising young programmer named Gus who I think would be great as my backup."

Sophie Jean leaned back.

Beethoven continued, "Gus needs a lot of training, but I can bring him up to speed."

She nodded and said, "That's a good idea, Beethoven. Have Gus see me in the morning." He got up and smiled over his shoulder at her, winking a knowing wink. She nodded back and said, "Yeah, you got me."

Beethoven walked through the office, glancing into the conference rooms filled with people talking and planning. He was proud to be an integral part of this effort—no, an indispensable part. At that moment, he felt confident. His countenance changed. His stride changed. His perception of the world changed. He grew up and focused outward. He became the person he had always envied. He'd always had the skills; he just lacked the confidence, but now he had both and had become the version of himself he had always hoped to be.

Moto saw him walk by and was struck with the change he saw. He wondered if he was as mature as Beethoven but figured it didn't matter as long as he was surrounded by mature people. Moto laughed out loud and went back to work.

***

Hurley headed up the team responsible for determining whether the backup environments had been compromised. Wires in 30 percent of the backup closets had patterns engrained in them near the outlets, but they were not patterns Hurley recognized. There were three distinct patterns: simple, complicated, and sophisticated. In the rest of the backup closets the wires were not affected.

Hurley surmised that 30 percent of the backups had been compromised three times, each time in a more sophisticated manner. He ran an analysis of the backups and found unauthorized access on another 23 percent of the backups - 53 percent had been compromised. He provided the list to Moto and the rest of the oversight team, along with the dates when the backups were accessed.

Moto told Lily he had to go manic to analyze the code on the compromised backups so Lily gathered Indy, Otis, Daisy Sue,

Beethoven, Branco, and the Gingers at Moto's apartment, in case Moto had questions, grabbing takeout on the way and arranging for more food in six hours. When she got to Moto's, she gave him a long, sloppy kiss - making that connection before he went manic. Soon, the others arrived and arranged the room with "cots" for everyone.

The cots contained a technology Moto designed when he'd had difficulty sleeping a few years before. The smart bed had a foam mattress with sensors that analyzed the nerve firings in the muscles and adjusted the firmness of the mattress in one inch square areas, adding and removing pressure to provide perfect comfort and support. The cot adjusted the head and neck to simulate a pillow. The sleeper had only to lie on the cot in any position, however bizarre, and in seconds the mattress adjusted to provide comfort and support. It also had a "huggy/snuggle" feature that made the sleeper feel like there was someone else in the bed. The cots were a little larger than a twin bed, but Moto's large living area could handle cots for the whole team.

Most had not seen Moto manic except for the brief period in the safe room a few days before. They had received communications from him during those times, but Moto spent those days cloistered at home. In the early years, he was ashamed of his bi-polar condition, but now that he could control it, he was proud of his superman ability and what he could accomplish in his manic state.

After greeting everyone and thanking them in advance for their help, Moto went to the corner with his computer, pictured his manic pattern, and focused. Lily told the others that Moto could not hear anything else going on in the room, but the team members couldn't bring themselves to talk out loud, so they whispered.

"Your whispers are bothering me," Moto announced unexpectedly, "Just talk normally," and he went back to work.

It was like a party. When they weren't fielding questions from Moto, they played chess, bocce ball (right-handed and left-handed), Family Feud, Clue, Cribbage, and HORSE on Moto's indoor basketball court. Not until the wee hours did the group finally retreat to their cots and drift off to sleep.

Lily touched Moto on the shoulder, and after a few minutes of internally organizing his analysis, he pictured his calming pattern and she to lead him to the bedroom for his required four hours of sleep. The house went dark and quiet, and Lily fell exhausted next to him, but in four hours, she was up, waking and feeding Moto so the cycle could resume.

The team didn't have Moto's ability to sleep at will or to wake up fully alert, but they did their best. Lily let most of them sleep an extra few hours, waking this person or that when Moto needed something.

As the day went on, Moto's questions became more pointed and detailed. After responding to each request, the team tried to figure out where Moto was heading, but they were always surprised at his next question.

Midday, Moto stood up and said he was going to take a shower, signaling he was finished, so the team left and Lily joined Moto in the shower.

Indy had been meditating in a closet and hadn't heard everyone leave, so when he came out it was to an empty room. He heard the shower but had something important to tell Moto, so he waited. He lay face down on the floor and covered his eyes in case Lily or Moto came out of the shower before getting dressed.

Lily spotted Indy when she walked into the living room wearing just a towel. She tip-toed back into the bedroom, and she and Moto dressed quietly. They came back out and Moto said loudly, "Oh, no. Here we are naked and Indy is still here. I hope he doesn't look up!" Then they jumped on him, tickling and hugging him playfully.

Between fits of laughter, Indy said, "Wait! Wait! I have something to tell you!" Lily and Moto sat back as Indy tried to catch his breath.

"Well, what is it?" Moto asked.

"I was meditating," Indy gasped, "and Bohdi came to me." Lily and Moto's eyes widened as Indy continued. "Her presence enveloped me. She didn't say any words, but I felt she was telling me 'you can do this.' I had a clear sense that she was telling us to be diligent, careful, and at the top of our game. Not that we *will* succeed, but that we can."

Lily's eyes stayed wide open, but Moto relaxed. "That's what we need to know," Moto said, "Thank Bodhi for me the next time you see her," Moto sighed and added, "I miss my grandfather. You're a lucky guy, Indy."

Indy nodded and when he left, Lily sat down next to Moto and gently leaned her head on his shoulder.

Chapter 45 – Moto Solves the Mystery

Chops sat in his office thinking about how the team had progressed in the last few years. He was glad to partially retire and spend time with Daisy but felt like he was abandoning them. He knew Branco would do a better job with the new challenges than he would, but he had been instrumental to this point. He would miss being in charge, but loved the feeling of being needed and loved by Daisy.

Daisy walked in and sat down, looking concerned. "I need to go to the east coast for a few days," she said, her voice shaking.

"Oh, why?" Chops asked, innocently.

"Obi is getting married this weekend and I'm his mother," she blurted out.

"Okay," Chops said, not looking up as if Daisy had said she was going to the grocery store.

"Chops, you already knew about Obi!" Daisy shouted, slapping him on the back of the head, "I'm sorry I didn't tell you he was my son before, but, uh…"

"I'm looking forward to introducing a seven foot Nigerian as my son – I'm going, too!" Chops laughed, and Daisy just shook her head.

"No harm, no foul," Chops said.

"No harm, no foul," Daisy echoed, kissing his ear.

Chops considered that if they had met earlier, he wouldn't have had time for a relationship. Daisy held his hand and they sat in silence except for the occasional giggle. "I'll get our plane tickets," Chops whispered and Daisy snuggled closer.

***

Sophie Jean called Riffy and Hurley into her office. She felt close to Hurley, but she wanted this meeting to be formal and professional.

Aloof and stand-offish, Riffy wasn't the type of person that Hurley would normally get along with - more the type of person that Hurley would chase out of a bar - but ever since the Maui trip they had been the best of friends. Hurley respected the danger that Riffy had willingly put himself in, and Riffy respected the work Hurley had done to keep him safe.

They walked into Sophie Jean's office, sat down, and said in unison, mimicking Sophie Jean, "I'll bet you're wondering why I called you in here." They waited a couple of seconds and continued, "Hurley has become too critical to the team and Riffy, you will be his backup—" Hurley couldn't finish the sentence because he broke out in laughter. Riffy at least finished the sentence before snickering.

Sophie Jean didn't break. "Go on," she said.

Hurley continued with the joke, but Riffy couldn't keep it up.

"This is a serious matter," Sophie Jean said sitting straight-faced and staring. Riffy and Hurley shifted uncomfortably. "I brought you two in here as a matter of extreme importance," Sophie Jean continued, her expression impenetrable, "but based on your attitude...." she couldn't hold it in any longer and giggled.

She jumped up, threw herself into Hurley's lap, put her arms around him, and kissed him sweetly.

Riffy scratched his head and, after a moment, said, "I guess I've been away a while." On his way out the door he added, "I'll be up to speed by the end of the week. I mean how much knowledge could Hurley have in his head?" but they didn't look up.

***

Edie shoved a large box into the bedroom of her new apartment. Although everything had been delivered, she still had a lot to do. It was hard to concentrate, as she hadn't heard what her new assignment was going to be. She'd returned a couple of days before and was getting worried. Had she done a good job? She went over the project and all the decisions she had made, second-guessing them all. She was emptying a box in the kitchen when Riffy walked up behind her.

"There's a new sheriff in town," Riffy said with a Texas drawl, standing back.

Edie spun around, her concerns melting away as she took in Riffy's smile.

"Congrats, Riff," Edie said busily. "That's perfect."

After a few seconds, "I looked at the personnel database for current and future assignments," Riffy said, while glancing at her unpacked boxes strewn about.

Edie waited for him to continue, but Riffy was silent, pretending to look at the boxes. She playfully whacked him and Riffy rubbed his head, "Just for that, you can wait until they tell you." Edie pinned him against the wall. "Okay, okay," Riffy said laughing, "but don't let on you know. You're going to be working with the Gingers to develop fake AI versions of all of us." It made perfect sense - Edie's experience with the AI bank manager code would be instrumental. "There's only one thing," Riffy added, pausing. Edie cocked her head in one direction, and when Riffy delayed further, cocked her head in the other direction, "You're going to be in charge of it!"

Edie smiled, her eyes sparkling, "Yeah, that's what I figured," she said, shrugging her shoulders, and as she bent down to pick up a box close to Riffy, she nudged his shoulder.

***

Lucy developed bronchitis on the way back from Europe and had been in bed for two days. Otis brought her chicken soup but rang the doorbell and left it outside, racing back to his car before she could answer. She didn't want to see anyone anyway, as she couldn't say two words without coughing. The meds and the chicken soup made her feel better, but she couldn't attend the safe room meeting and infect the team.

When she heard a knock on the door, she snapped, "Just leave it, Otis."

"It's Indy," she heard through the door. "Let me in."

Lucy opened the door and was surprised with a big hug. Indy took her by the shoulders and said, "My grandmother insists you go to the safe room meeting. You'll be well by then."

"But—"

Indy stopped her and held one of her hands in both of his. Lucy felt a chill run down her body. "You'll be fine," Indy said, turning around and he walked out the door.

The chill remained as she called Sophie Jean to let her know she would be attending. She had twelve minutes. The shower washed away the chill and she hurriedly dressed, noticing she felt better and hadn't coughed since Indy hugged her.

***

Moto was first to the safe room. Standing alone, he thought about the first meeting all those years ago, remembering how he didn't have a robe to wear because they were one short and how loyal everyone was to the cause. As his team walked in one by one, he gave each one a hug and thanked them. He could not have gotten this far alone. Humbled, appreciative, but also troubled, he had to tell the team what he had found. The anguish was overwhelming.

Sophie Jean began the meeting by expressing her gratitude for everyone's hard work. She reviewed the successes of past years and gave a shout out to each member for their contribution, forgetting no one. But there was no softening for what Moto was about to say.

"We've done a great job," she said, "but we're entering a new phase that will be much more challenging. It is far from a done deal that we will be successful. It will take innovation, execution, and smart work. It will take focus and coordination and the ability to react with precision. It will take sacrifice, determination, good health and a few lucky breaks—"

She paused to let it sink in. The faces in the room looked curious, but none showed discouragement or doubt. She concluded with "This fight will be winner-take-all, and it's one we have to win," sitting down, she motioned to Moto.

Moto stood up shakily, starting with "How do you like the new robes?" Moto had ordered custom robes, tailored to their bodies in colors that made them look their best. Some rolled their eyes, some nodded, but they were focused on what Moto had to say.

"Okay, here's the deal," Moto said. Then he stopped and looked around the room. He flashed back to the original eleven members of the core team, sitting in the same seats they had occupied at that first safe-room meeting. But the team had grown with Riffy, Beethoven, Edie, Branco, Weetzie, and Daisy Sue—all talented and motivated additions that he couldn't imagine being without. He took a deep breath and continued.

"Two and a half years ago there was a solar storm," Moto said. "It affected the electrical grid in Indonesia. Based on our analysis, it also affected one of our Sniffer programs, changing one bit of one program that was searching the grid for devices. That one small change prevented the program from shutting down, and it has been running on the grid ever since." Moto paused, gathering himself, then continued, "Our code updates modify programs in storage and those executing. These updates modified the rogue Sniffer program to attack the shut-down command rather than just ignore it. It targeted our shutdown command in order to stop it from being issued and searched our backups, to upgrade itself with various routines from our library. I call the rogue Sniffer program AIT1 for AI Threat number 1."

As Moto looked around the room, one by one, he saw the change in expression as they realized what had happened.

Moto's voice shook as he said, "I killed Shorty."

He lowered himself into his chair, and Sophie Jean put a steadying hand on his shoulder. She waited for the emotions in the room to run their course before she said, "The odds that the solar storm and the program updates would create this were one in 147 trillion, but the program code *we* wrote has produced our very first AI threat."

Sophie Jean glanced at Indy, who stood up, an imposing figure, drawing the room's attention. "My grandmother, Bodhi," he said, "came to me in a dream. She told me we could beat this," Indy paused, "*Could*, not would," he added sternly. "I know you're in shock, we've faced difficulties before and we have always found a path to success. If we build the right plan, we *can* be successful."

Indy sat down as Branco stood up. A few team members looked to Chops, but Chops looked intently at Branco and when the whole team turned their eyes to him, the transition from Chops to Branco was complete.

Before stepping through Moto's analysis, Branco said, "Moto reconstructed 85 percent of what AIT1 should look like. After all, it's our code that makes up the rogue program and that's our advantage, but there are unknowns. The wrong move could prove fatal, not just for us but for humanity. AIT1 has already shown it has a weapon that can kill with barely a trace, so it's urgent that we prevent another strike."

Moto informed the team that he was going to prepare additional insights and left the room, beckoning Lily to accompany him. As they walked arm in arm out the door, Lily leaned in close, sensing Moto was not comfortable being alone, recalling several times as she was leaving his apartment, he gave her a lingering look and asked her to stay. Sophie Jean had thought that it might be that Moto had reached a new level of love for her, but for now, Lily would stay with Moto 24x7.

They entered a small anterior safe room that Moto had recently added. "What do you need, Moto?" Lily asked.

"Just a pen and paper for now," he replied, "I have many insights into AIT1 that go beyond what Branco will present. I want to prepare more detail." Lily found a pen and a ream of paper, and Moto got to work.

***

As the meeting drew to a close, Moto and Lily returned with handouts —nine pages for everyone and another four to five individualized for each team member. Moloana, Indy, and Branco got a copy of everything. Once they reviewed Moto's additional insights, it prompted additional discussion and risk planning. The biggest risk was that Moto was not sure what code might have been incorporated into AIT1 or how it might have been modified.

Moto summarized, "Team, this rogue program is dangerous. It has incorporated the code we wrote into its own programming, but it is careful. Its primary mission is to keep running in an undetected mode, so if we try to send it an update with a virus or a fix, it may seek us out as an enemy like it did with the shutdown command. It uses our standard stealth technology to keep running by storing backups in various places. If a power grid containing the main program goes down, the communications end, the main program stops running, and the timer programs launch a new version of the main program."

"Do you already have a plan for how to deal with AIT1?" Edie blurted out without thinking.

"I'm sure he does, Edie, but to validate Moto's plan, we have to prepare an independent option to make sure we have the best possible solution," Branco responded quickly. "We can't do this without Moto, but he can't do it without us."

Sophie Jean looked around the room and then nodded appreciatively to Branco saying, "I have a feeling the team is ready to analyze and validate the strategy you're about to present, Moto." A chuckle rippled around the room, acknowledging that many had had the same thought as Edie. Sophie Jean demonstrated once again that she was the glue that held this team together.

Moto chuckled along with everyone else. "I guess you know me pretty well, Sophie Jean. I do have a strategy, but I'm guessing it is just a summary of what you've already discussed with Branco." Moto paused looking around the room at his friends before he continued. "The timers communicate with the main program every second, and they communicate with each other using a randomly generated pattern sent from the main program to ensure that all of the timers are working properly. If a timer fails to respond or pass a signal on to another timer, all the timers and the main program are notified, the backup copy of the main program associated with the unresponsive timer is wiped out, and a new timer and backup copy are identified." Moto suppressed a wry smile and added, "I'm sure this is all much too technical for you." The room shouted, almost in unison, "Hah!"

Convinced everyone was fully engaged, Moto continued, "The strategy with the least risk has three steps." Moto counted them off on his fingers as he spoke, "Step one is to find AIT1. Step 2 is to intercept the communication from the main program to the timers as it passes through the grid. The challenge is that the communication protocols have been modified, and it's the communications that killed Shorty. AIT1 combined our own protocols with our chemical fusion project to generate a micro fusion reaction to boost the power for communicating, which is what killed Shorty. AIT1 is bypassing our normal inter-grid communications."

Moto paused before continuing, but forgot to put his third finger up, "Step three is to obtain a copy of the AIT1 program code from one of the timers. If we can analyze the program code, we can eliminate AIT1 permanently."

Branco stood up and looked around the room, running a hand through his white hair. "Do we have any changes to our plan?"

Moloana raised her hand. "I have three new risk assessments." She went through the subtle differences of what Moto had presented and what Branco and the team had come up using insight that Moto had come to expect from her.

They had their plan and assignments. It was almost midnight, but no one went home. The next scheduled hardware check of the backups was in nineteen days. They didn't dare change the backup schedule or AIT1 might react in an unexpected way and they couldn't afford to have AIT1 obtain any more program code from the backups.

Chapter 46 – Indy Saves the World

Beethoven was under pressure to find AIT1. Indy dropped by and asked how he could help, but Beethoven wasn't sure. He was just getting started.

"Have you looked at where AIT1 has been?" Indy asked, trying to be helpful.

"Yes, but I can't see any pattern to it," Beethoven replied, frustrated.

Indy motioned at the doorway for Tilly to come in. "This is Tilly, Beethoven," Indy said, "Moto's swim coach."

"Okaaaaay," Beethoven said looking quizzically at Indy, "Nice to meet you."

"Let me see the data," Tilly snapped, leaving Beethoven with his mouth open.

"Tilly's good at patterns," Indy explained. "If anyone can find a pattern in AIT1's movements, Tilly can."

Beethoven nodded and gave her the tracking information. She examined it for a couple of minutes before she said, "You're missing data points. See if you can find a trace of AIT1 on grid 14372 in Malaysia and 15100199 in Seattle." Beethoven looked at Indy, who only shrugged, so he ran a search and surprisingly found traces left behind by AIT1 on those grids.

Tilly wrote down nineteen grid numbers and handed them to Beethoven, saying, "This is where AIT1 will be in the next fourteen hours. I'll be right back with its locations for the next seven days," and she turned quickly, walking out of the room.

"Glad I could help," Indy said with a wink as he left Beethoven with his mouth gaping.

Beethoven put programs in place to detect AIT1's presence in the grids Tilly provided and contacted Otis to tell him that locating AIT1 was imminent.

***

Otis froze - his team wasn't ready to try to intercept communications yet. He started to call Moto, but realized that without program specs Moto couldn't help yet.

Indy stuck his head into Otis' office. "I jotted down some ideas for the communication intercept," Indy said calmly, "maybe you can use them." Indy handed him a thumb drive and stepped out to tell Lily, who was walking by, to assure Moto got some rest, as they'd need him soon.

When Indy popped back in, Otis looked at him with disbelief. "These are completed specs for the communication intercept," he said, "When did you do this?"

Indy shrugged. Otis couldn't close his mouth. "When your team finishes validating these," Indy said, "ship them over to me. I'll take them to Moto," and he walked away leaving Otis staring at the empty doorway.

***

Branco was in his office updating the plan with dependencies and dates for the timeline. Indy loped in and stood next to his desk with his hands in his pockets. "How are things going?" he asked nonchalantly.

"I'm pretty nervous about this timeline," Branco replied.

Pointing to several items on the plan, Indy said, "These items are complete and you'll need to pull Moto in later today."

Branco looked closely and then updated the status, saying, "This new timeline will require Moto to be manic an inordinate amount of time. It's too risky for him to be without rest that long."

Indy looked over Branco's shoulder at the updated plan. "You're right, I'll fix that," he said walking quickly out of the room.

***

Tilly didn't have an office, only a cubicle, so Indy leaned over the back of her chair, whispered in her ear, and raced down the hall before she could protest.

***

When he reached the Gingers, Indy handed Ginger a thumb drive.

"What's this?" she asked.

"Specs for targeting a timer program to find AIT1's backup program code," Indy replied, "the backup of the AIT1 program should be in executable format so you will need to decompile the quantum executable code into programmable code so that Moto can analyze it," but before Ginger could respond, Indy was summoned into Sophie Jean's office where Tilly was waiting.

Otis glanced into the window of Sophie Jean's office and saw Indy standing, facing Tilly and Sophie Jean, arms folded and shaking his head, resolutely standing his ground to the animated pleadings from the two ladies. Finally, Otis saw Sophie Jean raise her hand and nod to Tilly in an obvious capitulation to Indy. Tilly ran out quickly and Indy bowed to Sophie Jean and left.

Indy arrived outside Moto's apartment an hour before Beethoven was due to provide the updated specifications. Lily came down to greet him, and having already heard what Indy had planned from Sophie Jean they discussed the logistics of Indy's plan.

When Tilly arrived, she stared at Indy as she walked up. "It is not ethical to put you in a manic state without a proven method to bring you back, Indy," she said, shaking. "I need to develop your calming pattern first or we may not be able to bring you back."

"The future of humanity is in jeopardy, Tilly. I need the manic pattern so I can write the code quickly. It's too dangerous for Moto – we need to save him to analyze the AIT1 program code. I can't do that analysis, but I can write the code we need now," Indy said firmly.

Lily looked at Tilly and nodded. It was clear that the risk that Indy was undertaking was less than having Moto in a manic state for ten days, so Tilly mouthed "OK" to Lily, who grabbed Indy by the hand and led him up to the apartment, asking "what are your food requirements."

"Burgers and steaks," Indy replied, not looking up.

Lily left to prepare Indy's food and Tilly put on her game face, "I'm not sure I can bring you back to a normal state, but I won't give up on you - you may need to rely on meditation." Tilly said, focusing on the notebook of papers she had laid out in front of her.

"Do you have a calming pattern for yourself?" Indy quipped.

Tilly couldn't smile and shook her head, "Let's get started."

Indy set up his workstation in a corner and checked that he had received the completed specifications. He nodded to Tilly, indicating that he was ready, and she showed him the first manic pattern.

Indy stared at the pattern and then back to Tilly, as he felt no change. She motioned with her eyes for him to look back at the pattern. He gazed and felt the pattern moving towards him, getting larger, enveloping him.

"I think it's working, but I'm not sure," he said.

"Focus on your workstation," Tilly said, but Indy could barely hear her, as if she were mouthing her words and he turned to his laptop with the program specifications on the screen. He could read all of the words on the first page at once with just a glance. He read the next page, then the next and the next, seeing all the words in his head.

"Now start coding," he heard Moto's voice say. Without looking up, Indy began writing code. He couldn't type fast enough to keep up with his mind. "Just picture the code on the page," he heard Moto say. Indy stopped trying to type with his fingers and visualized what he wanted to type, and the code appeared on the page. He had typed it, but without physical constraints.

Tilly's eyes widened when she saw how fast Indy was typing. Clearly, the pattern had worked, but Indy was sweating profusely.

Lily returned with Indy's food. "He's already started?" Lily said with some alarm. "I wanted to be sure he ate first."

"Oh, no," Tilly cried out, almost in tears, "I forgot!"

Lily ran to Tilly, "It's OK – I'll set up the IV – he looks like he is sweating – and I'll add enough nutrients to get him through."

Tilly calmed, but said abruptly, "I have to work on Indy's calming pattern," and turned away.

***

When Moto woke up the next morning, Lily was beside him. But before he opened his eyes she whispered, "We have company." Moto put on a robe, and shuffled into the living room, where he saw Indy flailing away at the keyboard and Tilly staring at several pieces of paper strewn around the floor. The memory of a dream he'd had the night before where he was teaching Indy how to swim slammed into his consciousness.

Moto strode back to the bedroom and shouted, "How could you let Indy do this?"

Lily stared at him fiercely, her eyes riveting. Several seconds went by. Moto was angry - Indy was risking his life and Lily was complicit.

"We're a team, Moto," Lily said at last, her voice quiet. "Indy understands the risk. You can't do it all — we'll need you later."

"Did Tilly develop his calming pattern?"

"She's working on it."

Moto's eyes looked down at his hands as if to say *'What have I done?'*

"He has a *decent* chance of coming back," Lily said, choking on the word 'decent'. "It was his choice. We're all in agreement"

Chapter 47 – Hurley Goes to the Big Island

Hurley's flight to the Big Island was bumpy and took forty-five minutes longer than scheduled. He hadn't told anyone he was going and he thought it might just be a fool's errand - but he had more than just a gut feeling.

When he arrived in Kona, he rented a car to go across the island to Hilo and after an endless drive, he turned north going another eighteen miles to a small, run-down house, where he stopped. He walked towards the front lanai just as a spry old man backed out of the door, still talking to someone inside. When the man turned around, Hurley's jaw dropped.

"Hey, you look kind of familiar," the man said. "Can we help you?" Hurley's mouth opened and closed again, but no words came out. "Well?" the old man said.

"Sorry, I guess I'm lost," Hurley lied. "I'm looking for Hilo."

The old man paused, and then shook his head saying "You want a beer? Then you can tell me why you're really here."

Hurley nodded and sat down on a rickety chair on the weathered lanai, rubbing his forehead.

When the old man came out and handed Hurley a Longboard, a woman in her fifties came with him. They sat for a while in silence drinking the beer.

"Well, son, you must want something to show up here," the man said at last, "not sure I can help you much, though. I've only been here a short time and don't remember much from before. Doc said I hit my head pretty hard." He turned, staring at Hurley, and added, "So what do you want?"

Hurley reached into his bag and pulled out a needle and a rubber strap. "Just your blood, sir," Hurley said smiling, and they broke into laughter. Hurley finished his beer, took a vial of the old man's blood, and said as he walked back to his car, "I'll be back." The old man nodded, holding the woman's hand as Hurley drove off.

Hurley put the blood in a cold pack he had brought and headed to Hilo to run the blood tests. Before he reached his hotel, he heard Sophie Jean say, "Taking a little vacation, Hurley?"

"Yep, always wanted to see an active volcano."

"Okay, I don't need to know?"

"Not yet."

Sophie Jean smiled and considered pulling the logs of what Hurley had been researching, but she was too busy with AIT1. She would know what Hurley was doing soon enough.

Hurley checked into his hotel and ate $46 worth of peanuts from the minibar while he started the blood analysis. Three hours later, the analysis was complete and the results were confirmed, but there was something else. Hurley stared at the results, at first mystified, then confused, and finally terrified. He ran the tests again and again with the same results.

Hurley grabbed the next flight back to the mainland.

Chapter 48 – Indy save Moto saves Indy

Otis received the first program from Indy three hours after sending the specifications and he assumed Moto was writing the code and Indy was the conduit. He passed it on to his team, and they set up a test environment, working around the clock to test the code, sending feedback to Indy. In three days, the program was ready.

Indy slumped in his chair in front of his workstation, his head shaking, and sweat rolling off his forehead. Tilly rushed to place the calming pattern she had developed in front of his eyes, but Indy didn't react. She moved the pattern closer to his face, but there was still no reaction.

"Indy, look at the pattern," she ordered, "Indy!"

Indy turned slowly toward Tilly and whispered, "I can't see it. I haven't been able to see anything for the last 24 hours."

Tilly gasped and leaned over him to read what he was typing: "Trace the calming pattern with my finger. I don't think I can speak any more and my hearing is getting weaker. I'm losing my sensory capabilities."

Tilly reached around Indy from behind, took his hands in hers, and used his fingers to trace the calming pattern. When she finished, she set his hands down, indicating the pattern was complete. Then she did it again, this time moving his hands in unison to show the three dimensions of the pattern. In reality, the calming pattern had five dimensions; the paper it was drawn on represented only two, but Tilly relied on the mind to translate the other three dimensions. Indy didn't react.

Panicked, Tilly dragged Indy onto on the floor, laying him on his back. She traced the five dimensions using his hands, feet, and head. She tried another calming pattern and another until she had none left.

Indy's physical responses slowed and in desperation, Tilly lay on the floor next to him, her arms and legs wrapped around him, holding him as tightly as she could. Indy's breathing became shallow, until he seemed to be taking his last breath. Overwhelmed with anger and frustration, Tilly gasped deeply and pulled away when she heard Indy say, "Don't go. I'm enjoying this."

Tilly looked at Indy's face and saw his open eyes and smile and she fell backwards.

Moto and Lily strode from the bedroom and Moto said, "I had the weirdest dream. What is everyone doing on the floor?"

Indy propped himself up on his elbows and said brightly, "Thanks for saving me, Moto. Lily, what is there to eat?"

Moto responded, "You mean my dream was real? I dreamt that you were zigzagging all over the sky, Indy, and that I flew up and pulled you back into my bed. It was a little uncomfortable with you, Lily, and me in the same bed, but you seemed grateful. I think it was Bohdi who told me to look up at the sky."

Indy smiled, "you have a very comfortable bed, Moto. Why am I lying here on the floor?" Tilly lay back in exhausted relief and laughed, and soon they were all laughing.

Lily went to the kitchen to get Indy a burger and Moto said wryly, "I guess you've made up for your sabbatical now."

"This going manic thing was quite an experience," Indy said, "I can see why it's useful to you, but I think going blind and deaf is a little much for me."

"Hmm, yeah, true that," Moto said grinning.

Indy stretched and yawned. "I guess I should check in with Otis and Beethoven to see how things are going," he said.

"If they need you, they'll contact you," Lily chirped, "Here's your burger," and she added, "You can use the bed in Moto's spare room. It's the same mattress Moto has. I'll wake you when you've slept long enough and you can be on your way."

Indy plopped down in front of the burgers and said, "Okay," yawning, "but I don't think I'll be able to sleep," and Moto smirked.

After eating, Lily led Indy to the soundproof guest bedroom and kissed him on the cheek. "Thank you, Indy," she said, "I love you." Indy mumbled something, lumbering to the bed and by the time Lily shut the door, he was asleep.

Chapter 49 - Making Progress

Otis and Beethoven established a war room to capture AIT1's program code. They adapted a program from the Jellyfish project they called the Interceptor, camouflaged to look like energy membranes in the grid, easily mistaken for normal fluctuations. Its purpose was to capture AIT1 communications and locate a backup copy of the program, while ensuring the messages were not delayed.

The Interceptors were launched successfully and captured the messages, giving Otis the format and content, but the communications were in pairs, indicating that AIT1 backups were stored in two pieces. To get a whole version of AIT1, they had to make some changes.

Otis gathered his team, quickly making the adjustments and Ginger stood by ready to convert the machine code into a more readable format. Otis pulled the switch again and the Interceptors were launched. In a few seconds, Otis received the executable programs (in two parts) and passed them to Ginger.

Ginger ran the modules through her "decompiler", but there were fatal errors; the program couldn't translate AIT1's code. It wasn't totally unexpected, but there was no time to fix it. She contacted Moto, who thanked her for trying but said he could read machine code almost as easily, so she sent him the modules.

Lily plied Moto with carbs and protein before he entered his manic world, where he was the center of his own universe and all around him was program code. He watched AIT1's program executing multiple simulations and sat back as it all played out.

***

Sophie Jean announced that Moto had begun the analysis to find AIT1's Achilles' heel and detailed what Indy had done to make it possible. "Wow" and "You're shitting me!" were the most common reactions, although a few people blurted out "WTF?"

Sophie Jean turned to planning another retreat for the team. She didn't want it to be a celebration but a gathering to bring the team closer together. After all, they were fighting an enemy they had created themselves, and they had lost Shorty, Max, and Brandy. She decided to take them back to Maui. Edie and Riffy were engaged now, as were Chops and Daisy, so a double wedding on the beach would be beautiful and romantic. She worked quickly as she feared that if she waited to start planning until this project was over, in all probability, a new crisis would block this well-deserved break.

***

Beethoven monitored AIT1's movements, confirming Tilly's projections of where the rogue program would go next, but as he received Sophie Jean's announcement, he suddenly lost track of AIT1. Panicked, he tried to contact Tilly – no response. Several minutes went by as Beethoven checked and double-checked the information Tilly provided. He stood up to run to Otis' office when he received a new list of projections from Tilly. He slammed into his chair and as he entered the new list into his tracking program, he felt a presence behind him.

"There were multiple power outages in the Ukraine and the Philippines," Tilly said close to his ear, "so I adjusted the future path of AIT1.   Stuff happens." She touched his shoulder and Beethoven saw that his tracking program had picked up the scent.

"I could tell you how to reprogram your tracking," Tilly continued, "but it would take too long and I couldn't be sure it would always work without my insight. Just be sure to call me if you lose AIT1 again and I'll update the pattern."

She's impressive, Beethoven thought, *what a unique talent – she's irreplaceable*. He thanked her profusely, causing her to walk away.

***

When Indy woke up, he was still tired. He lumbered out to the living room, where a perky Lily ran up and gave him a hug. Moto was typing at lightning speed with his left hand on his keyboard and gave Indy a shaka with his right without looking up.

Indy poured himself a cup of coffee and leaned across the counter to Lily "I want to stay with Moto to learn more about how he manages his mania. I want to know how he functions so well."

Lily looked at him incredulously, "You almost died. Don't forget that Tilly said never never picture that pattern again - it wasn't quite right - obviously."

"Yeah, going blind, not able to speak and dying - not cool."

Lily pointed to breakfast, gave Indy some instructions for Moto, and left to do errands, thankful for a break from tending to Moto.

After running around town for a while, she stopped by Sophie Jean's office to discuss Moto's anxiety, but when she walked in, she felt a strong need to be with him and turned to leave when Sophie Jean looked up.

 "Hi, Lily," Sophie Jean said, stopping her, "where's your sidekick?" Lily forced a smile but then broke into tears. Sophie Jean jumped up, led her to a chair, and sat next to her. Lily chuckled as she wiped her eyes. "I don't know why I did that," she said. "I came here to talk about Moto, but maybe we should talk about me instead."

"You're exhausted," Sophie Jean said, patting Lily's hand.

"Yes, it's really intense being with Moto all the time. I feel like I need time to myself, but just now, when I finally had time, I just wanted to be with him."

Sophie Jean smiled and said, "I'm so happy for you!"

"Huh?"

"Lily, your feelings of being overwhelmed by Moto when you're with him and your own separation anxiety when you're not are normal for a caregiver." Sophie Jean paused to let it sink in before continuing, "A loving relationship requires give and take. When one person cannot give equally, the other becomes the caregiver. This happens periodically in a normal relationship, such as when someone is sick, but it usually lasts for only a short time. People who have lost loved ones remember caregiving with great fondness, a time of closeness and feeling needed, even though it is physically and emotionally demanding. You're doing everything right, Lily. This is temporary - it won't be like this forever."

Lily looked down at her hands folded in her lap. Sophie Jean went on, "After this project, we'll discuss *your* needs. It will never be a 50-50 relationship - now it is 95-5 - but if it's 60-40, it will be workable."

"So I need to come to terms with owning the lion's share of the responsibility for our relationship?"

"Yes, but don't worry, you both love each other so much that it just has to work out."

Lily took a deep breath. "Thank you," she said with a smile. "You've helped me realize that having a relationship with Moto will always be a challenge, but I'm all in, and it's my responsibility to make it work."

Sophie Jean nodded and smiled, "Moto has responsibilities, too," she said softly.

Lily hugged her before striding down the hall to her office.

***

Indy sat comfortably in Moto's plush easy chair and set an alarm for his next Moto duty. Still whipped from his own manic episode, he feared he would fall asleep—which he did. In his dream, Indy saw Bodhi smiling at him and pointing off in the distance. He turned his head and looked: It was Moto, floating in the clouds. Indy turned back to Bodhi, who stared at him.

She drew a pattern in the clouds similar to, but not the same, as the one Tilly had given him. Suddenly, Indy was sitting on a cloud next to Moto, surrounded by executing program code, but Moto was shivering and panting. Indy reached out and put his hand on Moto's shoulder, and heard Moto's voice in his head asking for help.

Indy nodded, and Moto grabbed his hand causing Indy's mind to explode, seeing what Moto saw. Moto had been trying to find a weakness in AIT1 but couldn't find one. He feared that AIT1 was invincible to any computer virus.

In the next instant, Indy was awake in the easy chair and Moto was standing over him smiling. "Thanks, man," Moto said giddily, dancing around the chair.

Moto stopped dancing and said, "Let's talk about how we can take down AIT1," and he launched into a monologue lasting several minutes.

Shoving Moto to the floor, Indy yelled, "Stop! Write it down!"

"You're right!" Moto said, "I'm wasting time," and he ran to his laptop, documenting his findings for the team.

When he finished, Moto slumped in his chair and Indy spun him around, rolling him to the dining table. "Eat! Calming pattern! Sleep," Indy commanded. Moto dutifully stuffed himself and trotted to his room to calm down and sleep it off.

"Good job," Indy heard Lily's voice whisper, "Moto's monitoring system activated when he went to bed, and he's in good shape. It'll be fourteen hours before he needs anything else, so you can relax."

Indy was too vested in defeating AIT1 to relax, but responded with "Take your time – we're good here," and he heard a heavy sigh and a few sniffles as Lily signed off.

Chapter 50 – AIT1

The next scheduled hardware check at the backup sites was in just twelve days. AIT1 would be fishing for new code again, so the deadline loomed. Tilly chose the three backup sites that AIT1 would most likely target and Edie and Lucy had been tasked with ensuring that none of those backups had viable code that AIT1 could use.

They had a staff of twenty-three to assist them, but because of the redundancies and self-checking built into the original programs, they were all having difficulty making the code fail.

Edie brought the team together and opened saying, "We are good at making programs work well, but not so experienced in making them fail, so we have to make programs work to perfection – but to a different end."

Everyone just stared.

Lucy chimed in. "Our mindset should be to make program enhancements that work well – not make it fail. Let's say the enhancement spits in your eye, but you'd better be damned sure it spits in the eye that's open. And if they're both closed, then hold one open and spit in it."

Everyone just stared.

Edie looked at Lucy, shaking her head before saying "Define a counterproductive enhancement with appropriate security."

Finally, the team understood that they would need "enhancements" that would succeed in elaborate misdeeds.

Now in their element, they were back on track, and Edie and Lucy reported to Branco that they would be ready on time.

* * *

Beethoven and Otis received Moto's analysis, but it was not what they expected, so they brought the Jellyfish coding team in to assist. The methodology was elegant in its simplicity but elaborate in its technology. The solution was to reverse the one bit of information the solar flare had corrupted. By adapting the jellyfish technology to the electrical grid, the program would float through AIT1 undetected while changing the program back to its original setting. A simple shutdown command could then be sent, causing it to end processing.

The coding frenzy went on for several days. The safe room became a makeshift barracks. Moloana and Ginger entertained with their Timpani performance one day, and Edie did her best impression of Marilyn Monroe singing "Diamonds Are a Girl's Best Friend", but the pressure left no one unaffected.

Moto and Indy helped out where they could, but didn't have specific tasks assigned. Looking around the large room, Moto visualized the programs they were writing—a jellyfish consisting of bits and bytes. As he walked through the room, the jellyfish floated and changed, just as Moto had visualized at Little Beach.

Indy reprogrammed the wall-sized monitor to display test status, representing AIT1 as the Black Knight floating through an electrical wire. When the jellyfish approached, it surrounded the Black Knight, and after passing through, became the White Knight and they had a jousting contest. If the White Knight won they would hear "ding"; if not, the Black Knight laughed maniacally. During early testing the annoying maniacal laughter was changed to a "bong", but nonetheless continued to be overly annoying.

Their biggest challenge was alignment. AIT1 and the Jellyfish program moved through the electrical grid in a three-dimensional, freeform manner. Moloana suggested that the Jellyfish surround AIT1 until it had multiple confirmations of the location of the bit that had to be changed and the team agreed and went to work. The new code was put in place and Indy announced that he was putting the sound effects back in. The team looked down at their keyboards and shook their heads.

Testing continued, but now they heard "Ding! Ding! Ding! Ding!" They ran 145 consecutive successful tests. There was just eighteen hours before the next backup hardware check.

Otis was in charge of deployment, and Beethoven was prepared to send the shutdown command manually if the Jellyfish program became disabled. Tilly provided the grid location and everyone started walking into the war room. Otis deployed the jellyfish by entering the grid location on the console, and the screen went blank, followed by a loud "ding." People were still talking and coming into the room, looking for a seat. Moto had just stepped in.

Everyone stopped, turned around, and looked at Otis.

Otis shrugged his shoulders, "It worked."

"That's it?" Edie asked. Otis shrugged again and nodded.

"I guess so," Moto said.

Everyone stood in shock before slowly meandering towards the door, disappointed at the anti-climax after weeks of hard work. Sophie Jean ran to the door, herding everyone back in and, gaining their attention, said, "Moto has something to say."

As one, they turned and faced Moto. Moto scanned the room and then leaped into the air, a fist striking at the ceiling, yelling, "YEAAAHHHHHHHH!!!" The room exploded with cheers and hugs as Moto ran high-fiving and low-fiving, thanking everyone for their efforts. Sophie Jean grabbed the team leads to distribute envelopes containing invitations to Chops and Daisy's wedding, Edie and Riffy's wedding, plane tickets, hotel reservations, and vouchers for ten days on Maui.

Sophie Jean stood smiling at the excitement rippling around the room when she felt a tap on her shoulder. When she turned around, Hurley dropped to one knee and held out a small pink velvet box containing a black diamond engagement ring. The room went silent as if waiting to exhale. Sophie Jean shrugged and said, "Yeah, I guess" before leaping onto Hurley, tackling him to the ground, and the room erupted with cheers.

Lily wound her way through the excitement to find Sophie Jean and as she hugged her, said way too emotionally, "I'm so happy for you."

"Lily," Sophie Jean whispered into her ear. "You don't need an outward commitment from Moto. Your relationship is beyond that." They continued to hug, Lily blinking back tears of joy or sadness, she couldn't say.

Chapter 51 – A Celebration

Sophie Jean arranged for a private plane to take Moto and Lily to Maui a day after everyone else would arrive, but the morning of the flight, Moto and Lily woke up facing each other and said in unison, "Bodhi is concerned." Gasping and laughing, there was no denying a sense of foreboding. "Maybe it's just to keep us on our toes," Moto said with a mind-clearing shake of his head.

They arrived at the terminal early and stood at the check-in desk waiting for an attendant when Moto felt a tap on his shoulder. He turned around, but no one was there. Then Lily, too, turned around and said, "What is it?"

"What?" Moto said.

"Didn't you tap me on the shoulder?"

Moto shook his head, his eyes wide.

"You, too?" Lily asked.

"Let's take a commercial flight," Moto whispered.

"No argument here."

They grabbed their bags and rushed out of the terminal, finding the last two seats on the last flight to Maui. As they waited at the gate, TV news reported that a private plane that had left for Maui had gone down somewhere in the Pacific Ocean, although the pilot had parachuted to safety and been rescued by a fishing boat.

"Are you sure you want to go?" Moto asked.

Lily felt a soft, warm breeze on the back of her neck and said, "Definitely."

They boarded, sitting in the comfortable first class section, but they were not comfortable. On edge from Bodhi's warning, they fidgeted nervously. Lily watched Moto picture his calming pattern and go to sleep and she was glad he could get some rest, but wished she had her own calming pattern.

Instead, she thought about everything she was looking forward to on Maui–seeing Honey, the weddings, the camaraderie, and of course some quality time with Moto. It took the edge off for a while, but there was an underlying anxiety for most of the flight. She looked forward to talking to Indy about Bohdi's warnings.

When they arrived on Maui, Indy was waiting at baggage claim. "How did you know to switch planes?" Indy blurted out, adding only as an afterthought, "Uh, aloha." When they told Indy what had happened, he whispered, "Let's get out of here. Tell anyone who asks about changing your flight that it's complicated and that you'll tell them later. You don't want to lie … or tell the truth."

Once in the car, Indy talked nonstop about the wedding and activities planned for the next few days. He said Sophie Jean was hosting a kickoff meeting in the morning, and their wake-up call was already scheduled so Indy would pick them up to take them to the meeting. When they got to the hotel, Lily went straight to bed while Moto headed to the beach for a swim.

Lily didn't stir until the phone rang and when she opened her eyes, she saw Moto looking at her with a weird smile, his head propped on one hand. He handed her a cup of coffee with double cream and no sugar, which Lily sipped while Moto continued to gaze at her. Finally, Moto said, "Shower?"

That she understood.

***

Honey rushed to make last-second preparations, as the whole team was coming to *her* doomsday site. She had a surprise for them, as she had built a wedding chapel on the site's fourteen wooded acres, designed specifically for a double wedding, with bridal entrances on either side. It was a Hawaiian wonderland of greenery and flowers with a retractable canopy in case of rain. She had provided Hawaiian "formal" attire for the whole team for the ceremony, including special items for the wedding couples. When she heard it was going to be a triple wedding, she took it in stride, building a third bridal entrance and ordering an additional haku for Sophie Jean and a wedding lei for Hurley.

Honey was asked to schedule the wedding services for later in the week, but she thought all three couples would enjoy having their service earlier, then spending a few days honeymooning on one of the another islands. She didn't tell anyone she had changed the dates as she felt it was her prerogative as wedding coordinator, doomsday site manager, *and* island host. She arranged a few days in Waikiki for Chops and Daisy, for Edie and Riffy, it was the Big Island, and Sophie Jean and Hurley would be headed to Kauai. She was proud of the work she had done and hoped Sophie Jean would be pleased as she paced nervously waiting for her guests to arrive.

When Sophie Jean walked in, her mood was not what Honey had expected - she was anxious since Hurley had flown to the Big Island the day before. He was expected to be on time, but Sophie Jean was worried. Honey used this opportunity to kill two birds with one stone. "Come with me," Honey said, holding her hand out. Reluctantly, Sophie Jean took her hand and walked with Honey along a path to the most beautiful sight Sophie Jean had ever seen. She couldn't decide whether to laugh or cry, but either one would have been for joy.

"Oh, Honey," she started, but she couldn't find words other than "thank you."

Honey briefed Sophie Jean on the wedding plans and told her that all she had to do was look pretty. Time passed quickly as they discussed the traditional Hawaiian ceremony —blowing the conch, the vows, exchange of leis—until they had to dash back to the doomsday site entrance to greet the arrivals.

Beethoven, Otis, Lucy, Weetzie, and Branco came first. Sophie Jean couldn't help noticing Honey wink at Beethoven as he walked up. Chops and Daisy showed up next, followed by Indy, Moto, and Lily. Next was a limo with Edie, Riffy, the Gingers, and Moloana. Remi, Henry, Daisy Sue, and Tilly jumped out of the last car, but there was still no sign of Hurley.

Sophie Jean sighed and turned to go back inside when she heard a horn honking from a distance. She looked back and saw a car speeding up the road, followed by a trail of dust, with Hurley's arm dangling out the driver's side. Relieved, she hurried inside.

Wearing a more relaxed smile, Sophie Jean introduced Chops, who presented a summary of what the team had accomplished, making sure to point out all the successes achieved along the way. Next, the Gingers reported that the AI saturation schedule had been pushed out another nineteen months based on their success with RICKY and several other development sites. It was good news, but it meant another nineteen months of hard work ahead. Moto lauded each person's contributions in a way that was so sincere and appreciative that it brought some team members to tears. He had learned humility along with a deep respect for his team. Moto called for a moment of silence for Shorty, Max, and Brandy.

Last on the agenda was Honey, who bounded up to the podium, giving a review of the site development, promising a tour of the facilities after the meeting. She glanced at Sophie Jean, who nodded knowingly.

"And....," Honey said, hardly containing herself, "you will all be attending a triple wedding service in our new chapel today. There is a change of clothes for everyone, including the brides and grooms, and everything you'll need to get ready." Honey clapped excitedly and was soon joined by the rest of the team.

Sophie Jean herded the brides and grooms to a short rehearsal, and Honey took everyone else on a tour of the facilities, showing them where they could find their change of clothes.

Moto and Lily stayed behind, as they had already seen the facility and were not part of the service. They sat quietly for a while until Lily reached for Moto's hand.

"Moto—"

Hurley burst into the room, sweat soaking his shirt. "Moto," he said, breathless, "you have to come with me - you too, Lily."

Moto looked up. "Where are we going?"

"Follow me!"

Hurley led them to a room at the back of the site and before Lily and Moto could sit down, he was pacing and talking.

"A few weeks ago," Hurley said nervously, "I was reviewing facial recognition hits from our monitoring system for the escapees from RICKY. But then I saw an unexpected face appear." Hurley stopped pacing and faced Moto. "It was your grandfather, Eli."

Moto's jaw dropped, but before he could speak, Hurley continued, talking quickly. "I went to the Big Island and found him. He said he had been rescued at sea but had no memory from before that. I got a blood sample, and it's a match."

Moto's eyes widened and Hurley's brow furled. Lily looked from one to the other and, trying to keep Moto from getting too excited, said, "But that's wonderful, isn't it? What's wrong?"

"There's a problem," Hurley explained. "His memory is being suppressed. I did additional tests on his blood and they showed that he isn't taking any medication, but one of the tests detected Titanium." Hurley gulped and continued, "I had the blood analyzed with an electron microscope. There's nanotechnology in his bloodstream. The technology is unknown." Hurley reached for a chair, and Lily helped him sit down.

"Where is he?" Moto said quietly.

"Just through that door."

Lily grabbed Moto's hand to keep him from rushing in. "Wait, Moto," she said urgently. "Let him finish."

His breathing restored, Hurley continued. "I reviewed Eli's medical history, and there is a hole in it. He visited a doctor five years ago, about the time he was diagnosed with his illness, but there is no trace—I mean *no* trace of that doctor or the clinic. And there's no trace of the illness in his body either. I mean, I don't think he was ever sick."

Moto wanted to hug his grandfather and make him remember him, but there was an unknown technology at play. "I want my grandfather back," Moto said firmly. "I don't care what it takes."

Hurley nodded, "He's been sedated, so he will sleep for a while," and Hurley opened the door so that Moto could see his grandfather.

"Thirty minutes!" came Honey's voice on their Collars. Moto lingered at the doorway before following Lily down the hall. Discretion was required, as they didn't know what the nanotechnology was capable of or why it was there.

The guests arrived one by one at the enchanted chapel that Honey had built. The chairs were ergonometric white leather from Denmark, where the guests could relax and enjoy the pageantry instead of sitting in uncomfortable rickety white wooden chairs. Two bridal entrances were on the outside with another in the middle, and the chairs were arranged so that everyone could easily swivel and see each bride enter. Above the chapel were two floating balconies where Moloana and Ginger sat, wearing long, flowing white dresses. Indy, who was, of course, an ordained minister, stood tall and majestic under the wedding canopy, wearing three long, colorful leis over a traditional Hawaiian robe.

Honey cued Moloana and Ginger, who played an upbeat version of "Despacito" on the timpani. Chops came up the center aisle, dancing to the beat. When he reached the front, the music changed to "I Shot the Sheriff," and Hurley came in the left entrance, striding along with the music until he reached his position next to Chops. Then Moloana and Ginger switched to "Help!" as Riffy danced up toward the canopy with everyone shouting, "Help! I need somebody. Help! Not just anybody. Heeeeeelp!"

When Riffy reached the front, the three grooms, Hurley and Riffy in black and Chops between them in white, wearing traditional green and white open wedding leis, grasped hands and raised them in the air, which lifted Chops a foot and a half off the ground.

Moloana and Ginger performed a timpani drum roll that reached a crescendo, before softening until there was silence. After a moment, Ginger began Lohengrin's "Wedding March," which Moloana accompanied with timpani drum rolls. The music filled the chapel with quiet majesty, surrounding the guests with a haunting, hypnotic air. The three brides entered: Daisy, all in white ruffles and a white Haku, gliding up the middle aisle to a beaming Chops; Edie, with her flowing curly red hair, bouncing toward Riffy in a long red dress and a red and white Haku; and Sophie Jean in a short white dress and a bright pink Haku drifted slowly and deliberately to Hurley.

The blowing of conch shells preceded the presentation of the leis and exchange of vows, Indy pronouncing them married.

Honey pressed a button. White plumeria petals showered down as the three couples stood with arms raised at the center of the chapel. Honey pressed another button and the chapel mechanically transformed into a dance floor, a buffet rising from beneath the altar. Moloana and Ginger played upbeat music and Moto walked around hugging people as if in a daze. He applauded Hurley. He clapped Riffy on the back. He picked Chops up and twirled him around.

Stepping away, Moto sat with Lily, holding her hand. His grandfather, the AI threat, the Jellyfish project, Bohdi's warnings, all the projects at Ball and Chain swirled violently in his head. He thought about Lily and his incredible team, neither of which he could do without. Neither of which he could let down.

Lily rubbed her thumb along the back of Moto's hand and whispered in his ear, "Let's go do some good."

Moto looked into her eyes, "I'm in."

www.ingramcontent.com/pod-product-compliance
Lightning Source LLC
Chambersburg PA
CBHW060304260626
47160CB00007B/2506